THE MAN I KNEW AS GRANDAD

The Soldier he never spoke about

DAWN HEMMINGS

WOODBRIDGE
PUBLISHERS

ISBN: 9798905143014 (Paperback)

ISBN: 979-8-90514-302-1 (Hardback)

ISBN: 979-8-90514-301-2 (eBook)

Cover Design by Woodbridge Publishers.

Find out more about our upcoming releases and authors at www.woodbridgepublishers.co.uk and sign up for our newsletter to stay updated!

DEDICATION

To my grandad,

***Arthur Hemmings** — my hero.*

And to my family,

so that you may know the man behind the silence.

CONTENTS

PREFACE

My grandad never told his story properly.

Not all at once. Not in detail. Only fragments.

A desert. An escape. A moment of such desperation that he drank his own urine to survive.

To me, he was just Grandad.

But to the wider world, he had been something else entirely, a soldier, a prisoner of war, a man who endured more than he ever chose to share.

As a child, I would sit and listen to his stories. They were never told like a history lesson, never neatly laid out or fully explained. They came in fragments, brief glimpses of places and events that felt distant and unreal.

He once spoke of escaping, though he never said exactly where, and of being so desperate that he had to drink his own urine to survive. As a child, I pictured him wandering across a desert, the sun relentless, sand stretching endlessly in every direction.

Only years later did I begin to understand what those fragments meant.

The desert that he spoke of was Tobruk. The escape was real, both there and later in Italy. The hardship was real. And the silence was real, too.

Like many men of his generation, he carried his experiences quietly. He rarely complained and rarely explained.

Silence was his armour, and we, his family, were left with only fragments.

Now, those fragments have taken on a different weight What began as a childhood curiosity has become something more: a responsibility to understand and preserve his story before it fades.

Many in my family know very little about the man he was during the war. They know him only as he was afterwards. But that part of his life shaped everything that followed.

And that matters.

Because the man I remember, the one with permanently oil-stained hands and a quiet, steady way about him, had lived through something extraordinary.

I began piecing his story together through letters, records, and the places he had been.

But reading about it was not enough.

I travelled to Italy. I walked the ground he walked, stood where he had stood behind barbed wire, and looked out across the same mountains he crossed during his escape.

Only then did it begin to feel real.

Where records exist, I have followed them closely. Where they do not, I have drawn carefully on context, place, and the fragments he left behind to understand what may have been.

This book is both a history and a family record. It follows my grandad from enlistment to captivity, from hunger and uncertainty to escape and, finally, freedom.

But more than that, it is about understanding the man behind the silence.

I want my family, especially the younger generations, to know who he really was. Not just a name in the family tree, but a man of resilience, quiet strength, and courage.

A man who endured.

I hope you read this with the same sense of wonder I felt as a child.

Only now, I understand what I was really hearing.

CHAPTER ONE

WHO WAS ARTHUR HEMMINGS?

Before the war, he was simply a man with oil on his hands and a family to provide for.

By the time it was over, he had been a prisoner, an escapee, twice, and a man who would spend the rest of his life saying very little about it.

This is where his story begins.

Arthur Hemmings was born on 21 April 1917 in Stockton-on-Tees. His mother, Eileen Mary Elgie, was just sixteen. His father, Herbert Hemmings, was nineteen.

They were young, raising a child in a world still marked by one war and moving steadily toward another.

Arthur grew up alongside his siblings, Harold, Dorothy, and June. Life was not easy, but it taught him how to manage, how to share, and how to endure.

At some point, the family moved to Scunthorpe, though the reason remains unclear.

By trade, Arthur was a mechanic. He worked with his hands, practical, capable hands, often marked by oil and grease that never fully washed away. It is one of the things I remember most about him.

He fixed things. He made do.

That skill stayed with him all his life and would later prove vital in circumstances far removed from any garage.

In 1936, at just nineteen, he married Elsie Parker. Together, they began building a family.

By the outbreak of war, they already had two young sons: Brian Herbert, born in 1936, and Geoffrey Arthur, born in 1939. They were too young to understand what was happening, but the war would shape their childhood all the same.

On 4 April 1940, Arthur enlisted in the British Army as a Private.

At the time, men between eighteen and forty-one were required to register, undergo a medical, and be sworn in. It was presented as a duty, but for many, it was simply inevitable.

Britain was already at war. Even where the bombings had not yet reached, there was no escaping it.

Posters appeared on walls "Your Country Needs You." The wireless carried speeches. Soldiers moved through the streets.

Arthur likely set out early that morning.

It is possible he did not walk alone. Perhaps Elsie went part of the way with him, Brian holding her hand, Geoffrey in his pram. Pride and fear would have sat side by side.

The enlistment office was likely the Territorial Army centre on Cottage Beck Road. The distance was not far, but the weight of the journey would have been.

Others would have been walking the same way. Some talking too loudly, masking nerves. Others silent.

Boots on stone. A steady sound.

Inside, the process was quick and impersonal. Names called. Papers signed. Men were examined, passed, or rejected.

Arthur passed easily. He was twenty-three, young and strong.

When he walked back out, he was still the same man but he no longer belonged only to his family.

He belonged to the Army now. To the war. To a future that would take him far from home.

CHAPTER TWO

NO 172388 PTE

A HEMMINGS MILITARY RECORD

Of all the documents I have found about my grandad's war, his military records read the coldest.

It is nothing more than dates, places, and numbers.

Periods of engagement. Days counted.

Movements logged.

Yet behind each line is a stretch of life lived under pressure, fear, hunger, and uncertainty.

When I read it slowly, those dry entries begin to breathe.

Arthur Hemmings enlisted on 4 April 1940.

For the first eighty-two days, he remained at home service in Glasgow, for training, before embarkation.

Those early months were about turning civilians into soldiers.

Long days. Cold mornings. Endless drills.

Learning to march, to follow orders, to become part of something larger than himself.

Scotland would have been a shock for a man used to England's industrial towns: warmer weather, wide spaces, and the growing sense that this was only the beginning.

On 25 August 1940, his records shifted to Egypt.

From training, he moved into active service overseas.

For one year and three hundred and sixty-one days, his life disappeared into the broad sweep of the war.

During this time, he served with the Royal Army Service Corps, later attached to the Auxiliary Military Pioneer Corps, eventually becoming part of 31 Company Army Motorised Petty Corps AMPC.

These were not glamorous units.

They were the backbone of the Army, the men who built, repaired, hauled, and kept everything moving.

To understand where Arthur was, you must understand why he ended up in Egypt.

British troops were not there by accident.

Egypt was vital to the war effort.

The Suez Canal was the lifeline of the British Empire, the fastest route for ships carrying troops, supplies, and oil from Asia and the Middle East.

If the Axis powers (mainly Germany and Italy) controlled it, Britain would be cut off from much of the world.

Under the Anglo-Egyptian Treaty of 1936, Britain was allowed to keep troops in Egypt specifically to defend the canal.

Although Egypt was officially neutral, it became one of the most important Allied bases (Britain, the USA, and the Soviet Union) of the entire war.

From there, British and Commonwealth forces launched operations across North Africa and into the Middle East.

Camps, depots, workshops, and airfields spread across the desert, and men like my grandad were sent there to keep the machinery of war moving.

Arthur was classified as a Fitter II.

This suited him, being a mechanic by trade.

The Army put that skill to use.

As a Fitter II, he was responsible for keeping vehicles and machinery running, often in rough conditions with limited tools and parts.

He repaired lorries, maintained engines, fixed pumps, winches, and plant machinery.

When something broke, he made it work again.

Sometimes that meant proper repairs in a workshop.

Often it meant improvising in the field, fabricating parts from whatever metal or scrap was available.

The Pioneer Corps might have been known for hard manual labour, but men like Arthur made sure the work could be done.

Sometime between May and June 1942, 31 Coy AMPC moved from Egypt to Tobruk because of the changing tactical situation on the Western Desert campaign.

This happened for a couple of reasons:

1. Germany and Italy advanced after the Gazala, when Axis forces broke through British lines, Allied forces were ordered forward for redeployment to hold key positions. Tobruk became critical as a defensive and supply base

2. Tobruk was a vital supply and logistics hub. The AMPC job was transport and supply as the front-line troops shifted west towards Tobruk. Therefore, supply units' like 31 Coy had to move with them to deliver fuel, food, and ammunition, evacuate casualties, and keep lines of communication running.

So, 31 Coy AMPC did not move to Tobruk by choice or coincidence; they were following the front-line and the supply chain as the British position in North Africa shifted westwards under pressure from Axis forces.

That long stretch of service ended abruptly on 20 June 1942.

Tobruk fell.

Rommel's forces broke through and captured the port; more than thirty-two thousand Allied soldiers were taken prisoner.

Arthur was one of them.

But he escaped from Tobruk.

For nine days.

Those nine days are some of the hardest to imagine.

The land around Tobruk was not rolling sand dunes, but hard, open dry rocky coastal desert.

Rock, dust, and scrub baked white under the sun.

There was nowhere to hide, no shade, no water.

He later told us he had escaped into the desert, that it was hot and empty, and that at one point he was so desperate he had to drink his own urine to survive.

As a child, I thought it sounded unreal.

As an adult, I understand it for what it was: a man pushed beyond dignity, beyond comfort, doing whatever it took to stay alive.

His military record was then changed to simply state: "Captured Tobruk."

No explanation. No emotion. Just a fact.

But I know from his letters, and from the fragments he shared years later, that what followed was chaos.

After nine days on the run, he was recaptured.

He became a prisoner of war.

In one of his first letters to my Nana, from Campo 65, he wrote,'

"I had a good try in getting away when Tobruk fell, it took them 9 days after before I was captured, and believe me, darling, what a thrilling experience, I thoroughly enjoyed every minute of it."

After capture and after he rejoined the other prisoners, he was loaded onto an Italian warship and taken to Italy across the Mediterranean Sea.

During my research, I have not been able to determine which port he arrived at, but it would have been either Naples on the west coast or Bari on the east coast.

Once in Italy, they were forced onto trains like cattle, crowded with no possibility of sitting.

The train headed east or west into an unfamiliar country.

His train stopped.

He had arrived at Campo 65, one of Italy's largest and harshest prisoner-of-war camps.

He would remain there until June 1943, living behind wire, enduring hunger, monotony, and uncertainty.

In June 1943, Campo 65 was closed, and prisoners were redistributed.

Arthur was transferred again, this time to Campo 145 at Poggio Cancelli, high in the mountains of central Italy.

From June until October 1943, he was held there.

In total, his time as a prisoner of war lasted one year and one hundred and twenty-nine days.

Then, in October 1943, he escaped again.

This time, it was different.

Italy had surrendered to the Allied forces in September 1943.

Guards had abandoned camps.

Authority had collapsed.

Arthur slipped away from Campo 145 and walked into the mountains. For three weeks, he moved through forests and ridges, hiding by day, walking by night, relying on instinct, luck, and the quiet courage of Italian civilians who risked their lives to help him.

Eventually, he encountered Allied forces advancing north.

The records note that he stumbled across the British North Africa Force and remained with them for forty-three days, until 9 December 1943.

After everything he had endured, he was finally out of danger.

On 10 December 1943, Arthur returned to the United Kingdom - Home.

At the top of his service record, one line stands out:

"*Repatriated POW ex Italy. Arrived in the UK 10/12/1943. Not to be sent overseas before 10/06/1944.*"

Six months' protection. A recognition, perhaps, that he had already given enough.

He never went overseas again.

From December 1943 until July 1946, he remained in home service, a further two years and two hundred and twenty-seven days.

The war ended.

Life began, slowly, to rebuild.

On 25 July 1946, he entered Class Z Reserve, officially released from active service.

He was still tied to the Army until February 1954, when he was finally discharged at the age of thirty-seven.

His name was then transferred to the Reserve List until he turned forty-five.

And that was it.

On paper, it is neat. Timed. Counted. Balanced.

But when I look at those dates together, what I see is a man who spent years moving from home to training, from desert to captivity, from wire to mountains, from fear to freedom.

A mechanic turned soldier. A prisoner turned escapee. A quiet man who survived when many did not.

The Army counted his days.

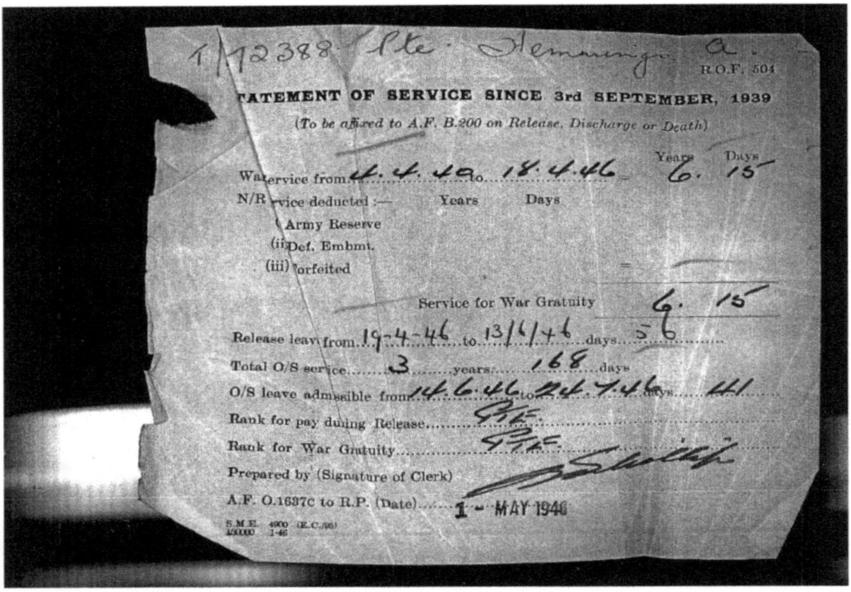

T/72388 Pte. Hemmings A.

R.O.F. 504

STATEMENT OF SERVICE SINCE 3rd SEPTEMBER, 1939

(To be affixed to A.F. B.200 on Release, Discharge or Death)

War service from 4.4.40 to 18.4.46 — Years 6. Days 15

N/R service deducted :— Years Days

(i) Army Reserve

(ii) Def. Embmt.

(iii) Forfeited

Service for War Gratuity 6. 15

Release leave from 19-4-46 to 13/6/46 days 56

Total O/S service 3 years 168 days

O/S leave admissible from 14.6.46 to 24.7.46 days 41

Rank for pay during Release Pte

Rank for War Gratuity Pte

Prepared by (Signature of Clerk)

A.F. O.1637c to R.P. (Date) 1 - MAY 1946

S.M.E. 4900 (E.C./95) 1-46

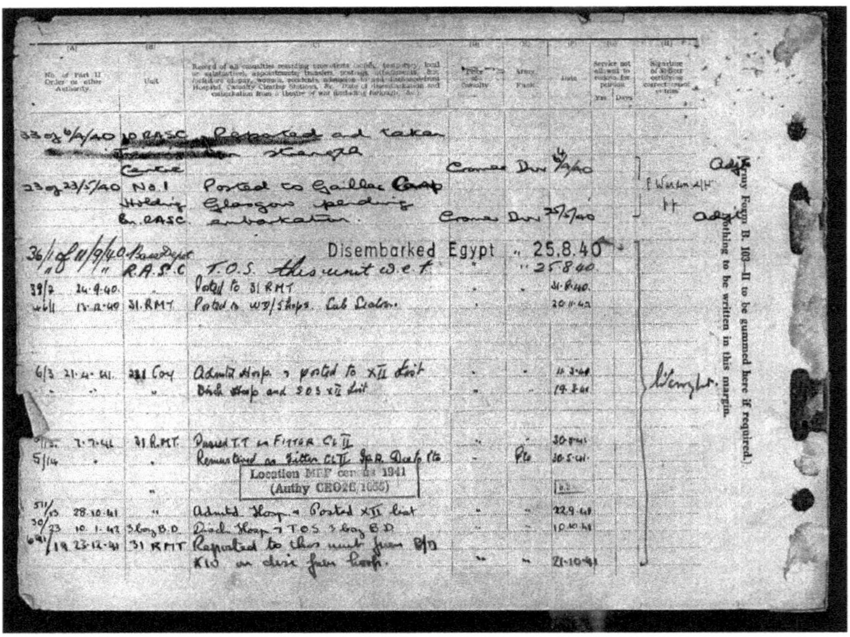

Disembarked Egypt 25.8.40

T.O.S. this unit w.e.f. 25.8.40

Posted to 31 RMT

Admitd Hosp. & posted to XII List

Disch Hosp and SOS XII List

Passed T.T. as Fitter Cl II

Remustered as Fitter Cl II

Location MEF census 1941 (Authy CRO2E/1055)

Admitd Hosp & Posted XII List

Disch Hosp & T.O.S. 3 Bay B.D.

Reported to this unit from B/D

Army Form B. 103—II to be gummed here if required.

Nothing to be written in this margin.

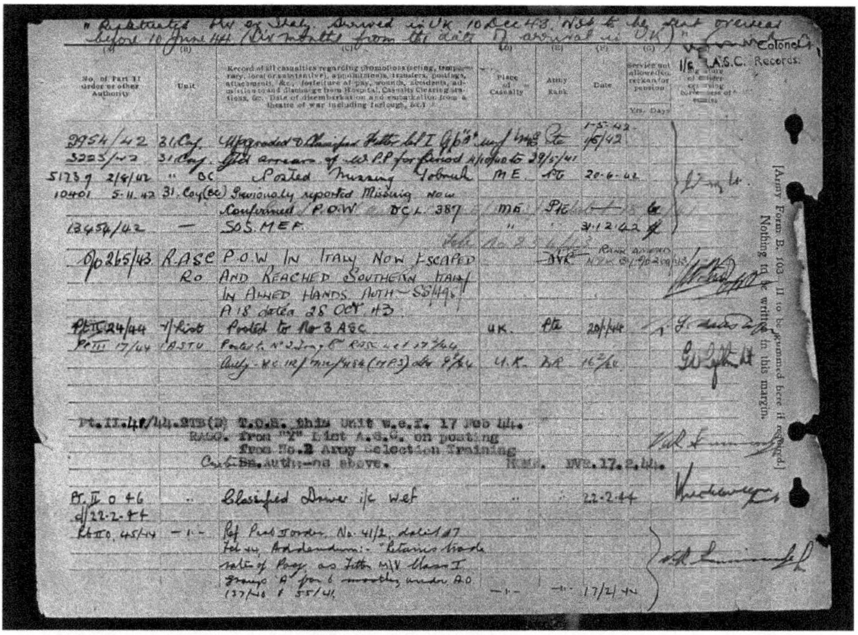

"Repatriated PW ex Italy. Arrived in UK 10 Dec 43. Not to be sent overseas before 10 June 44 (Six months from the date of arrival in UK)" Colonel i/c R.A.S.C. Records.

No. of Part II Order or other Authority	Unit	Record of all casualties regarding promotions (acting, temporary, local or substantive), appointments, transfers, postings, attachments, &c., forfeiture of pay, wounds, accidents, admission to and discharge from Hospital, Casualty Clearing Stations, &c. Date of disembarkation and embarkation from & theatre of war including furlough, &c.	Place of Casualty	Army Rank	Date	Service not allowed to reckon for pension Yrs. Days	Signature of officer certifying correctness of entries
2954/42	31 Coy.	Upgraded & Classified Fitter Cl I Gp "A"		Pte	1-5-42 16/42		
3225/42	31 Coy.	Granted arrears of W.P.P. for period 4/10/40 to 29/5/41					
51739 2/8/42	" BC	Posted Missing Tobruk	M.E.	Pte	20-6-42		
10401 5-11-42	31 Coy (BC)	Previously reported Missing now Confirmed P.O.W. DCL 387	M.E.	Pte			
13454/42	—	S.O.S. M.E.F.	"		31-12-42		
O/0265/43	RASC Ro	P.O.W. in Italy now escaped and reached Southern Italy in Allied hands. Auth – SS/495 A18 dated 28 Oct 43.		Dvr			
Pt II 24/44	Y List	Posted to No 3 ASC	U.K.	Pte	20/1/44		
Pt III 17/44	ASTU						
			U.K.	Dr	16/2/44		
Pt.II.41/44.2TB(D)	RASC	T.O.S. this Unit w.e.f. 17 Feb 44. from "Y" List A.S.C. on posting from No.2 Army Selection Training Centre Auth:–as above.					
Pt. II 0 46 d/22-2-44	..	Classified Driver i/c wef	..		22-2-44		
Pt II 0. 45/44	–"–	Ref Part II Order No. 41/2, dated 17 Feb 44, Addendum:– "Retains trade rate of Pay as Fitter M/V Class I Group A for 6 months, under A.O. 137/40 & 55/41.	–"–	–"–	17/2/44		

[Army Form B. 103. Nothing is to be written in this margin. II to be gummed here if required.]

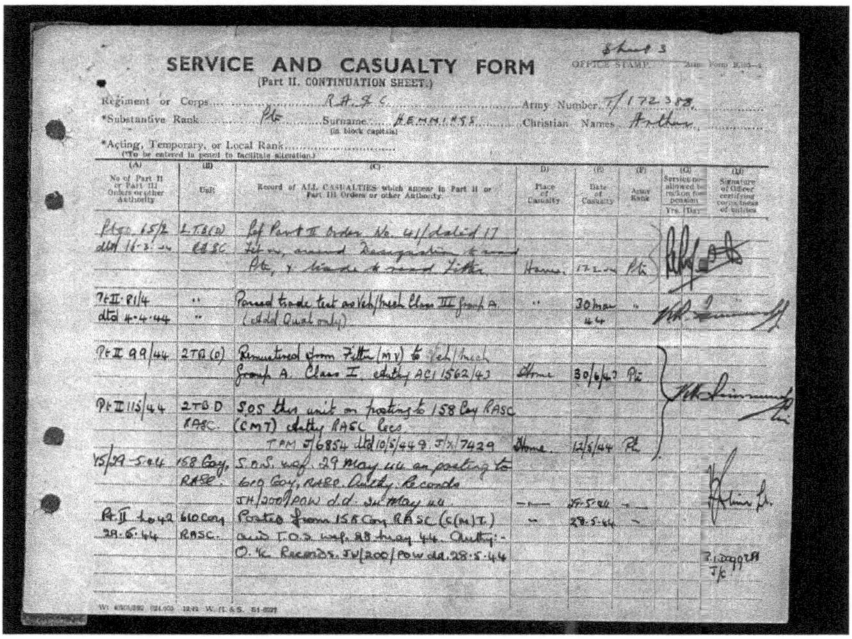

Sheet 3

SERVICE AND CASUALTY FORM

(Part II. CONTINUATION SHEET.)

OFFICE STAMP.

Regiment or Corps R.A.S.C. Army Number T/172388.

*Substantive Rank Pte Surname HEMMINGS (in block capitals) Christian Names Arthur

*Acting, Temporary, or Local Rank (To be entered in pencil to facilitate alteration.)

(A) No. of Part II or Part III Orders or other Authority	(B) Unit	(C) Record of ALL CASUALTIES which appear in Part II or Part III Orders or other Authority	(D) Place of Casualty	(E) Date of Casualty	(F) Army Rank	(G) Service not allowed to reckon for pension Yrs. Days	(H) Signature of Officer certifying correctness of entries
Pt II 15/2 dtd 16-3-44	2 T.B.(D) RASC	Ref Part II Order No. 41/dated 17 Feb 44, amend Designation to read Pte, & trade to read Fitter	Home	17-2-44	Pte		
Pt II 81/4 dtd 4-4-44	.. ..	Passed trade test as Veh/Mech Class III Group A. (Add Qual only)	..	30 Mar 44	..		
Pt II 99/44	2 TB (D)	Remustered from Fitter (MV) to Veh/Mech Group A. Class I. Authy ACI 1562/43	Home	30/4/44	Pte		
Pt II 115/44	2 TB D RASC	S.O.S. this unit on posting to 158 Coy RASC (CMT) Authy RASC Recs					
		TAM 7/6854 dtd 10/5/44 9. Jx/7429	Home	12/5/44	Pte		
15/29-5-44	158 Coy. RASC	S.O.S. wef. 29 May 44 on posting to 610 Coy, RASC. Authy Records JH/200/POW d.d. 26 May 44	—	29-5-44	—		
Pt II 1042 29-5-44	610 Coy RASC.	Posted from 158 Coy RASC (C(M)T.) and T.O.S. wef. 28 May 44. Authy:– O.C. Records. JV/200/POW dd. 28-5-44	—	28-5-44	—		

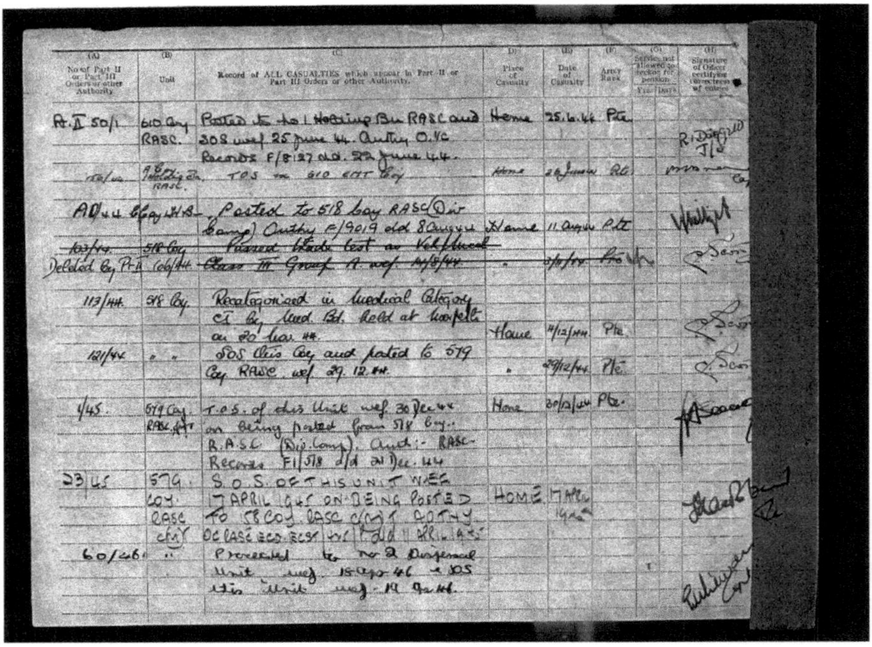

(A) No. of Part II or Part III Orders or other Authority	(B) Unit	(C) Record of ALL CASUALTIES which appear in Part II or Part III Orders or other Authority.	(D) Place of Casualty	(E) Date of Casualty	(F) Army Rank	(G) Service not allowed to reckon for pension — Yrs.	Days	(H) Signature of Officer certifying correctness of entries
Pt. II 50/1	610 Coy RASC.	Posted to No 1 Holding Bn RASC and SOS wef 25 June 44. Authy O.i/c Records F/8127 dd. 22 June 44.	Home	25.6.44	Pte.			R. [illegible] J/C
[illegible]	[illegible] Holding Bn. RASC.	TOS on 610 [illegible] Coy	Home	26 June 44	Pte.			[illegible] Capt
AD/44	[illegible]	Posted to 518 Coy RASC (Div Camp) Authy F/9019 dd 8 August 44	Home	11 Aug 44	Pte			[illegible]
~~103/44.~~ Deleted by Pt II 106/44.	~~518 Coy~~	~~Passed trade test as Veh/Mech Class III Group A. wef. 14/8/44.~~	"	~~3/10/44~~	~~Pte.~~			[illegible]
113/44.	518 Coy.	Recategorised in Medical Category C1 by Med. Bd. held at [illegible] on 20 Nov. 44.	Home	4/12/44.	Pte.			[illegible]
121/44	" "	SOS this Coy and posted to 579 Coy RASC. wef. 29.12.44.	"	29/12/44.	Pte.			[illegible]
1/45.	579 Coy. RASC [illegible]	T.O.S. of this Unit wef. 30 Dec 44 on being posted from 518 Coy. R.A.S.C. (Div. Camp). Auth:- RASC Records F1/518 d/d 21 Dec. 44	Home	30/12/44	Pte.			[illegible]
23/45	579 COY. RASC C(M)	S.O.S. OF THIS UNIT W.E.F. 17 APRIL 1945 ON BEING POSTED TO 58 COY. RASC C(M) AUTHY. OC RASC [illegible] d/d 11 APRIL 1945	HOME	17 APRIL 1945				[illegible]
60/46	"	Presented to No 2 Dispersal Unit wef. 18 Apr 46 & SOS this Unit wef. 19 Apr 46.						[illegible] Capt

CHAPTER THREE

HISTORY OF TOBRUK – 1942

Tobruk was more than a port. It was a symbol.

Sitting on Libya's Mediterranean coast, its deep natural harbour made it one of the most valuable positions in North Africa.

In 1941, Allied forces had held it against repeated Axis assaults during a long and punishing siege.

The men who defended it became known simply as the "Rats of Tobruk," a name first intended as an insult by the enemy, but later worn with pride.

They had survived constant bombing, equipment and food shortages, and isolation.

Tobruk stood for endurance.

But war is never still.

By June 1942, the balance of power in North Africa had shifted.

Field Marshal Erwin Rommel and the German Afrika Korps launched a fierce and coordinated offensive.

Allied lines collapsed under pressure.

Communications failed.

Units were cut off from one another. What had once been a stronghold quickly became a trap.

On 20 June 1942, Tobruk fell.

More than 32,000 Allied soldiers were captured in a matter of days.

It was one of the largest surrenders of British-led forces in the war.

For the Allied forces at Tobruk, there was little time to prepare, little time to think.

Confusion ruled. Some surrendered. Some were surrounded. Others ran.

My grandad ran. He escaped for 9 days.

Once back in custody, he joined thousands of other Allied prisoners gathered in and around Tobruk.

British, Australian, New Zealand, and South African men who only days earlier had been soldiers defending a position, now reduced to numbers in enemy hands.

Some prisoners were taken to POW camps at Tobruk under guard while transport was organised.

Conditions were harsh. Food was scarce. Water is tightly rationed.

The desert heat did not soften simply because they were no longer free.

Disease was a constant threat. Guards were stretched thin, and discipline varied, depending on who was in charge.

Uncertainty hung in the air heavier than the dust.

But for my grandad, the long journey north began.

Prisoners were marched and then loaded onto ships bound for Italy.

The Mediterranean was dangerous, patrolled by submarines and aircraft.

Below deck, conditions were cramped and airless.

Men lay shoulder to shoulder in crowded holds, already weakened from battle and escape.

Seasickness mixed with dehydration. The voyage was another trial layered on top of the last.

When they reached Italy, processing camps awaited them.

Names were recorded. Numbers assigned.

Identity shifted from soldier to prisoner of war.

From there, my grandad and many others were transported inland by train.

Wooden carriages, packed tight. Little ventilation.

The journey was slow and stifling.

Outside, Italy rolled past, olive groves, farmland, distant hills, landscapes of beauty and normal life.

Inside, men sat in sweat and silence, unsure where they were going or how long captivity would last.

Eventually, the train pulled into the Apulia, also known as Puglia region of southern Italy, directly in front of the camp.

Campo 65.

One of Italy's largest prisoner-of-war camps, designed to hold thousands, twelve thousand to be precise.

Long stone barracks.

Barbed wire fences.

Eight watchtowers rising at intervals. Order imposed by routine.

Roll calls. Rations. Waiting.

The desert had tested my grandad, alone.

Campo 65 would test him differently, through monotony, hunger, confinement, and the slow erosion of time.

Here, survival meant patience. Discipline. Mental endurance.

Looking back now, I see Tobruk not as a single event but as the turning point.

It was a moment when my grandad's war changed.

Before it, he was a soldier serving abroad.

After it, he was a prisoner navigating captivity.

Those nine days in the desert were the beginning of a pattern that followed him through the rest of the war, escape, survival, and silence.

He would escape again in Italy. He would walk through mountains as he had crossed the desert. He would endure.

As a child, I heard only the extraordinary story. As an adult, I understand the cost.

He spoke calmly about things that should have broken him. He gave us fragments, not details. Enough to know he had suffered.

Enough to know he had survived.

That memory has stayed with me.

ROMMEL TAKES TOBRUK
IN SWIFT DESERT BLITZ

New Aleutian Isle Taken

Japanese Land On Kiska; U. S.

The Honolulu Advertiser

FIRST EDITION

Weary Allies Are Crushed By Nazi Army

Bundesarchiv, Bild 101I-786-0301-39
Foto: Dörner | Juni 1942 ca.

German troops make their way into the burning port of Tobruk.

Bundesarchiv, Bild 101I-786-0307-12
Foto: Moosmüller | 1942 August - September

German infantry moving into Tobruk.

Bundesarchiv, Bild 101I-785-0299-22A
Foto: Moosmüller | Juni 1942 ca.

Rommel and Bayerlein survey the harbour at Tobruk, still crowded with bombed out ships that had accumulated in over a year's fighting.

Bundesarchiv, Bild 101I-785-0294-32A
Foto: Tannenberg | Juni 1942 ca.

Some of the 33,000 Allied Prisoners of War are marched away from Tobruk.

CHAPTER FOUR

MY JOURNEY TO ITALY

I didn't set out planning a trip to Italy.

At first, all I wanted to do was understand my grandad a little better.

I had his letters, carefully kept, written in his neat handwriting, each one carrying a camp address stamped across the top. Campo 65. Campo 145.

Names that meant nothing to me at the time, just places on paper linked to a part of his life he rarely spoke about.

When I decided to write about him, I knew I couldn't rely only on memory.

Stories fade. Details blur. I needed facts. I needed context.

I needed to know how he became a prisoner of war and where those places actually were.

So, I started researching.

I typed the camp names into search engines, half-expecting to find very little.

Instead, I found maps.

Dozens of camps scattered across Italy, far more than I had ever realised. Approx. 76 official POW camps. In addition to the main camps, there were hundreds of smaller agricultural and labour sub-camps attached to them.

I had heard about the German camps but never the Italian.

They weren't hidden away in one area but spread across the country: coastlines, valleys, farmland, mountains.

It shocked me how many there were, and how little I had known about them.

Those addresses from his letters suddenly had geography. Distance. Meaning.

During my research, I came across a website dedicated entirely to Campo 65, the first camp my grandad had been held in.

It wasn't an official government site or a polished museum page.

It felt personal, careful, respectful. It was clear that whoever had created it cared deeply about the place and its history.

On impulse, I sent an email.

I waited.

I was expecting perhaps a brief reply, or none at all.

Instead, I heard back from a man named Domenico Bolognese.

His message was warm, kind, and generous.

He told me about the Association Campo 65, a group working to preserve the camp and its history, not just for visitors, but for memory itself.

He offered to show me around the site personally if I ever came to Italy.

That was the moment things changed.

What had begun as research slowly turned into something else.

An idea. A pull.

I realised that reading about the camps was not enough.

My grandad had been there. He had walked those grounds, slept in those huts, stared out at those landscapes.

If I wanted to understand him properly, I needed to go too.

So, I planned the trip.

Standing at Campo 65 with Domenico was an experience I hadn't prepared myself for.

He didn't rush.

He spoke quietly, carefully, explaining what the camp had been, how it functioned, and what life there had been like for the men held inside.

He talked about the importance of preserving not just buildings, but stories. Of making sure places like this were not erased by time or development.

At the end of the tour, he handed me a book the association had written: *Camp 65 – The Memory That Remains.*

It felt like a gift far greater than its pages.

It was proof that people who had never met my grandad still cared about what had happened to him and to thousands like him.

Domenico also pointed me toward another resource, one that helped me learn more about Campo 145, Montorio al Vomano, province of Teramo, the mountain camp where my grandad would later escape.

Slowly, the pieces began to connect.

The letters. The maps. The places.

The journey he had been forced to make all those years before.

By then, Italy no longer felt distant. It felt personal.

What struck me most was how ordinary everything looked now.

Fields. Hills. Quiet villages.

Places where life had moved on.

And yet, beneath that calm surface, history waited.

Not loudly. Not dramatically.

Just patiently, for someone to ask the right questions.

Visiting Italy was never about ticking off locations, like I normally do.

It was about standing where my grandad once stood and realising that his story was real, physical, grounded in soil and stone.

It was about understanding that his survival wasn't just something that happened to him. It shaped all of us who came after.

By the time I left, I knew one thing for certain.

This wasn't just research anymore.

It was a journey I needed to make, to honour him, and to bring his story home.

CHAPTER FIVE

CAMPO 65 – ALTAMURA GRAVINA

Situated between Altamura and Gravina in Puglia on the S5596, is situated Campo 65.

It was the largest prisoner-of-war camp in Italy.

On paper, to the officials who built and ran it, it was nothing more than a grid of stone huts, a perimeter of wire fences, a scattering of eight watchtowers, and a few buildings for officers.

To anyone looking at it on a map, it was a neat arrangement of rectangles in an empty stretch of southern Italian countryside.

But to the men who were detained there, it was something else entirely.

They called it by many names in letters, diaries, and quiet conversations, but the one that lingered, the one that captured its essence most honestly, was "Hell on Earth."

My Grandad was held here from June 1942, after his capture at Tobruk, to July 1943, when it was closed.

The camp was built to hold non-commissioned officers and enlisted troops, the rank and file of armies captured in North Africa and transported across the Mediterranean.

At its peak, Campo 65 imprisoned up to twelve thousand men.

Imagine that number for a moment.

An entire town's worth of soldiers crammed into long, narrow barracks, each one a stone hut with walls that seemed to sweat in the damp, and floors that turned quickly to mud beneath the constant shuffle of boots.

The air was thick with mildew.

At night, condensation dripped from the ceilings onto blankets already sodden from humidity.

Mosquitoes rose in clouds from the swamp that pressed close to the camp, bringing with them the feverish spectre of malaria.

Diphtheria stalked through the huts too, carried in coughs, fevers, and the quiet groans of men whose strength was already stripped thin by hunger.

Unlike other camps hidden in forests or nestled in mountain valleys, Campo 65 stood exposed.

The land around it was bare and open, a flat plain broken only by distant hills.

Miles of wheat fields and fallow ground stretched in every direction.

For men who dreamed of escape, it offered no shelter.

Even if you managed to slip past the watchtowers and barbed wire, there was nowhere to vanish to. No cover, no villages within easy reach, no sympathetic locals who might smuggle food or hide you in barns.

And if you were caught, if the guards dragged you back, the punishments were brutal.

Cells meant for three men would be stuffed with thirty. Beatings were routine, torture not unheard of. To men already weakened by sickness and starvation, such punishments made survival feel almost impossible.

Food was always the enemy.

Not the guards, not even the war itself—at least not directly.

Hunger was what gnawed at the spirit day after day. The official rations were poor: a crust of bread, a ladle of watery soup, sometimes a few beans floating in a broth that smelled faintly of cabbage or onions.

Never enough.

Men grew obsessed with food, talking about it, dreaming about it, trading scraps and favours for an extra bite.

Now and then, prisoners were sent out to work on a nearby farm, and those jobs were sought-after, not because the work was easier, but because there might be a crust of extra bread, a handful of olives, the corner of a cheese rind.

Campo 65 operated only for a short window, from April 1942 until June 1943. Barely a year.

Yet in that time, it carved a scar deep enough that men remembered it for decades.

After the war, when testimonies were collected, letters resurfaced, and survivors spoke, a flood of complaints poured out about the conditions.

Investigations were launched into whether the camp's treatment of prisoners crossed into war crimes.

Later, under Allied control, Campo 65 was repurposed as a refugee camp, holding displaced families long after the guns had fallen silent. It remained in use until 1962.

But nothing that came later, not the new fences, not the refugees' laughter, not the passing of years, could erase its wartime history.

I wasn't prepared for how I felt as I walked through the gates of Campo 65.

I had spent time researching the camp.

The drive there had been uneventful, the kind of countryside you almost get used to after hours on the road, rolling fields, patches of scrubland, the odd farmhouse standing stubborn against the wind.

If you were not looking for it, you would pass by without even noticing.

Most of the site cannot be seen from the road. The only thing you would glimpse, if you happened to glance in the right direction, was the officers' quarters, a solid square building of stone, still standing with a strange sort of pride. A fountain situated in the front.

You would never guess that just beyond the hill behind them sprawled the remains of one of Italy's largest wartime camps.

I arrived early and sat and waited for Domenicio.

I stood there looking at the entrance to the camp; it felt surreal to be there.

But as soon as we stepped through the gate, silence closed in.

Not ordinary silence, like the kind you notice on a country walk when the wind drops and birds pause their song, but a heavier silence.

It pressed into me, as though the air itself was thickened by memory. It was the silence of a place that had seen suffering, where the ground itself seemed unwilling to speak too loudly.

I had come with a purpose.

Most of the barracks are long gone; only two remain.

There is nothing left of the others but foundations and the faint outline of a floor plan.

But I could picture it: rows of wooden bunks stacked three high, straw mattresses blackened by damp, men pressed together under lice-ridden blankets, their breath clouding in the cold air or overheating from the heat.

The smell of sickness, the constant coughing, the restless shifting of men too weak to sleep soundly.

My grandad's letters home were guarded, clipped, and carefully worded, but the address was there, scrawled at the top of the page: Campo 65, Prigionieri di Guerra, Baracca 16.

From my grandad's letters, I knew he had been kept in Camp 4, Building 16.

Domenico and I looked at a photograph of the camp and managed to identify exactly which building it was.

As I stood on the foundations of building 16, it sent a shiver through me.

And in that silence, I began to imagine him there.

I imagined at night, in the dim light of the barracks, men whispered to each other. The guards never cared about whispers; it was shouts they feared.

"Where d'you think we are, lad?" one voice might ask, hushed but urgent.

"Near Bari, I heard," another would reply. "Baron land all around. No getting out of here."

My grandad, I thought, might have kept quiet at first, listening.

He was that sort of man, patient, steady. Then maybe he spoke softly, not to argue, not to give false hope, but to anchor them.

"Doesn't matter where we are. We just have to get through it. One day at a time."

A cough would break the stillness.

A groan from another bunk.

Hunger and fever didn't stop for conversation. But those few words, steady, calm, quiet, might have been the rope men clung to in the dark.

By day, he might have been sent to the farm behind the camp, where groups of prisoners worked under guard.

I pictured him with sleeves rolled up, shoulders bent over a hoe, the sun beating down, dust clinging to sweat.

His hands, already roughened by mechanic work, now blistered anew under the weight of forced labour.

Perhaps he shared a rare laugh with another prisoner when the guards weren't looking.

"Funny, isn't it?" his mate might have muttered, wiping his brow. "Digging up their fields to feed them, while we're starving inside."

Grandad may have given a half-smile, that quiet sort of humour that doesn't burst into laughter but lingers like a stubborn ember.

"You have to laugh, don't you? Otherwise, you'd go mad."

Those fragments of imagined conversation, their ordinary quality, felt truer than any official report.

Starvation, disease, and punishment: yes, those defined the camp. But so too did the whispers, the muttered jokes, the fragments of humanity men held onto in order not to vanish inside themselves.

Walking through the remains, I felt goosebumps rise along my arms.

The officers' quarters looked much as they must have eighty years ago, the stone fountain still intact.

And the watchtowers, all eight of them, still stood in their places, mute sentinels over an empty field. They had watched then, and they watched now, their windows blank but somehow still menacing.

I paused by one tower, looking up, imagining what it must have been like for him.

Each day, the guards peered down from above, rifles slung casually across their shoulders.

To the prisoners, they would have seemed distant but absolute, sometimes cruel, sometimes indifferent, always in control.

Beyond them stretched the fields, empty and unreachable. Freedom visible, yet impossibly far.

Standing there, I thought of the strength it must have taken simply to endure.

To wake each day into hunger, sickness, monotony, and fear.

To watch others collapse and be carried out yet keep going.

In his letters, Grandad never once described the details of his suffering. He wrote instead about the weather, about missing home, about small things that reassured his family he was alive.

That silence, his silence, was a shield.

I wondered if, inside the camp, he was the one who steadied others. The one who reminded them that the war would end, that life waited beyond the fences, that survival was enough.

Campo 65 today is still a quiet, lonely place.

The land around it has hardly changed, the same rolling fields, the same pale sky.

From the outside, it looks ordinary.

Step inside, though, and you feel its weight. The silence carries stories, even if no one is speaking.

Stories of fear, of endurance, of men who clung to scraps of hope when everything seemed lost.

For me, Campo 65 is no longer just a site on a map, no longer just a name in a history book. It is part of my family's story.

To know that my grandad walked here, worked here, suffered and survived here, it reshapes the way I see the place, the way I carry his memory.

And standing in that silence, I could almost hear his voice in the still air: steady, quiet, reassuring.

"One day at a time."

Gate to Campo 65

Map of Campo 65

In living and everlasting memory of the Istrian, Fiume, and Dalmatian exiles forced to abandon their homes after the Second World War and who found refuge in this camp.
- Fiumani Studies Society, Fiume Historical Museum Archive (Rome), Altamura, October 6, 2023
A VIVO E PERENNE
RICORDO
DEGLI ESULI ISTRIANI,
FIUMANI E DALMATI
COSTRETTI
AD ABBANDONARE
LE PROPRIE TERRE
DOPO
IL SECONDO CONFLITTO
MONDIALE
E CHE
IN QUESTO CAMPO
TROVARONO ACCOGLIENZA
SOCIETA' STUDI FIUMANI
ARCHIVIO MUSEO STORICO DI FIUME (ROMA)
ALTAMURA, 6/10/2023

Rows of buildings behind the barbed wire

Campo 65 Barracks

Campo 65 Barracks original layout

Inside one of the remaining blocks

The Italian Officers Quarters along with a fountain at the front

View of Campo 65 from the road

One of the remaining huts

The only drawing of what the rooms in the blocks looked like

Inside one of the blocks

My Grandads letters back where they were written

CHAPTER SIX

LETTERS FROM CAMP 65

During my grandad's time in the Italian prisoner-of-war camps, he wrote regularly to my nana.

I inherited these letters from my dad.

Some of them make me smile, others make me stop and think, but all of them offer a glimpse into what life was really like behind the wire.

They are not dramatic or self-pitying.

Instead, they are practical, sometimes gently humorous, and often carefully restrained.

Between the lines, you can sense what was happening in the camps, what he chose to share, and what he kept to himself.

They all started with, '*Dearest Elsie, just a few lines to let you know that everything is alright.*'

I want to share those letters with you, beginning with Campo 65.

They begin with the telegram my nana received, the moment her life and his, changed, telling her that my grandad was now a prisoner of war.

All letters have been typed because the handwriting is difficult to read, in places.

Although my Grandad only had two children, he will refer to his two sons as kiddies followed by and not forgetting Hazel. Hazel had no family, so my Great Grandma took her in, and she ended up living with my Nana and Grandad.

For these letters to reach the UK, each followed a long route, travelling from Rome to Lisbon before finally arriving in London and then onto Scunthorpe.

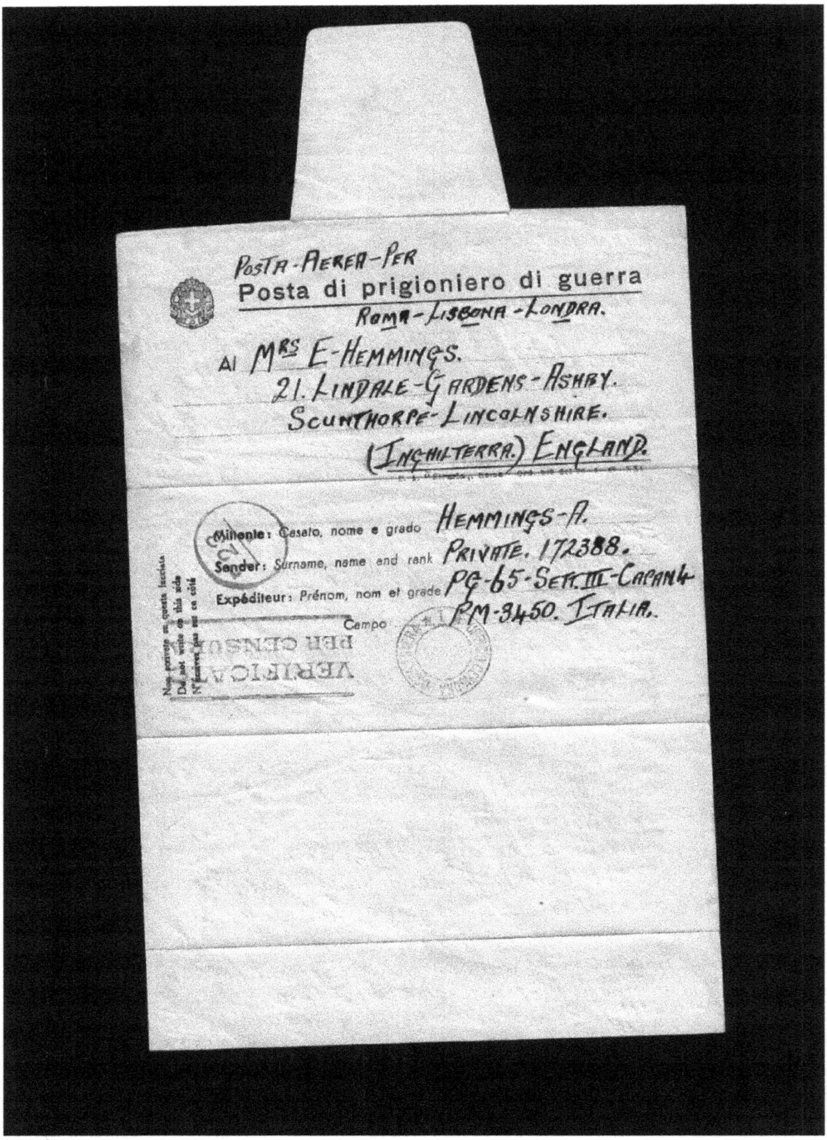
POSTA-AEREA-PER
Posta di prigioniero di guerra
ROMA-LISBONA-LONDRA.

Al MRS E-HEMMINGS.
21. LINDALE-GARDENS-ASHBY.
SCUNTHORPE-LINCOLNSHIRE.
(INGHILTERRA) ENGLAND.

Mittente: Casato, nome e grado HEMMINGS-A.
Sender: Surname, name and rank PRIVATE. 172388.
Expéditeur: Prénom, nom et grade PG-65-SETT.III-CAPAN4
Campo PM-3450. ITALIA.

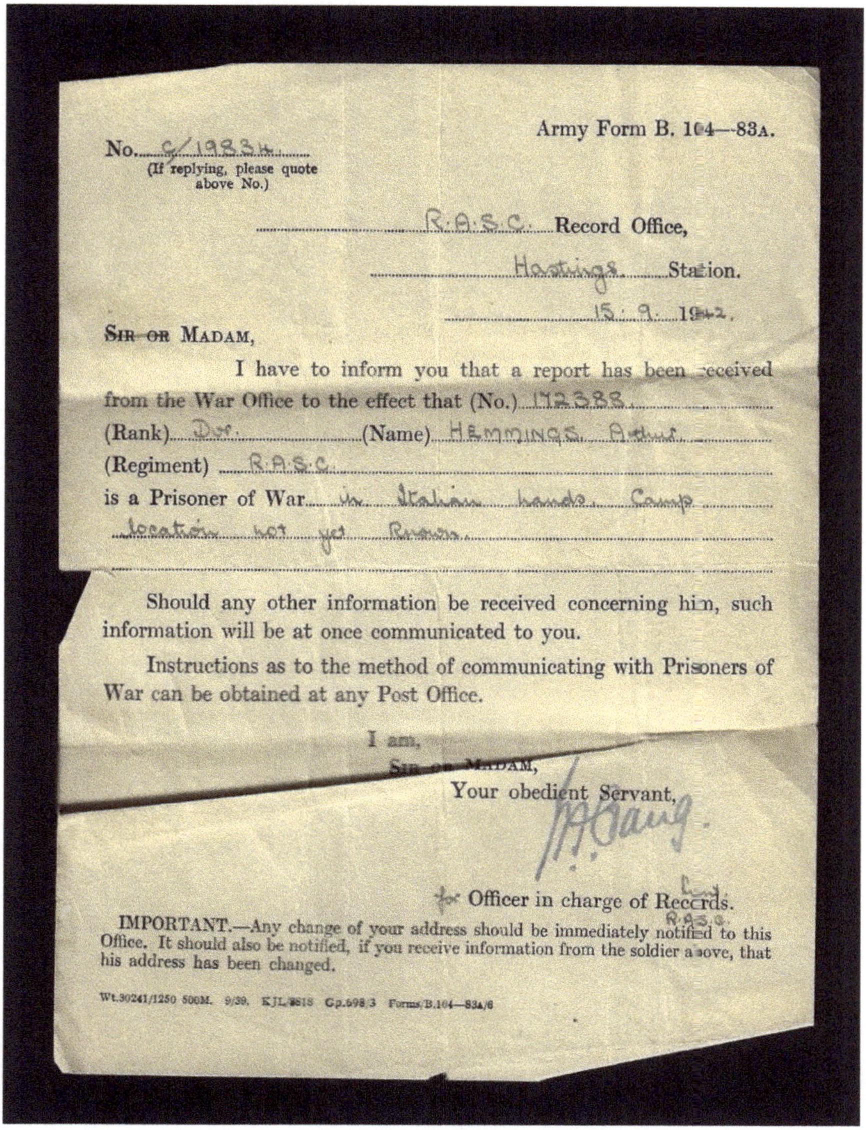

Army Form B. 104—83A.

No. C/19834.
(If replying, please quote above No.)

R.A.S.C. Record Office,

Hastings Station.

15. 9. 1942.

~~Sir or~~ Madam,

I have to inform you that a report has been received from the War Office to the effect that (No.) 172388.
(Rank) Dvr. (Name) HEMMINGS, Arthur
(Regiment) R.A.S.C.
is a Prisoner of War in Italian hands. Camp location not yet known.

Should any other information be received concerning him, such information will be at once communicated to you.

Instructions as to the method of communicating with Prisoners of War can be obtained at any Post Office.

I am,
~~Sir or~~ Madam,
Your obedient Servant,
[signature]
for Officer in charge of Records. R.A.S.C.

IMPORTANT.—Any change of your address should be immediately notified to this Office. It should also be notified, if you receive information from the soldier above, that his address has been changed.

Wt.30241/1250 500M. 9/39. KJL/8618 Gp.698/3 Forms/B.104—83A/6

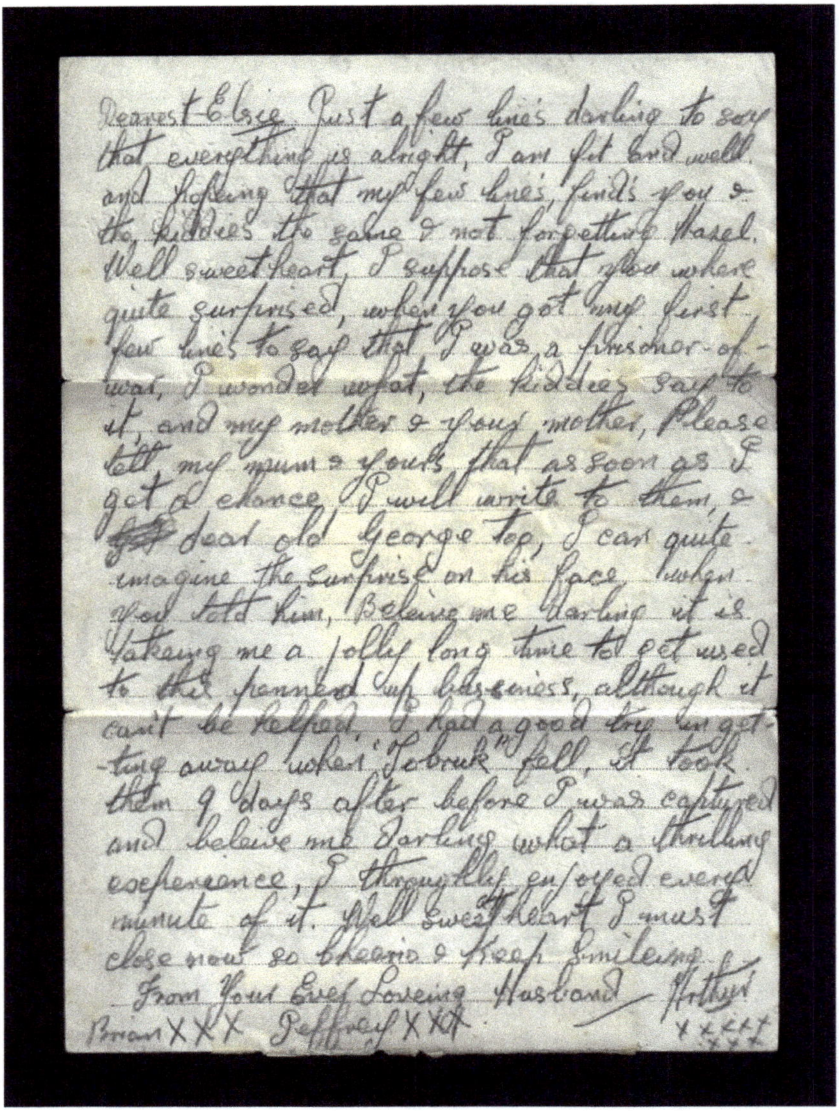

Dearest Elsie Just a few line's darling to say that everything is alright, I am fit and well and hopeing that my few line's, finds you & the kiddies the same & not forgetting Hazel. Well sweetheart, I suppose that you where quite surprised, when you got my first few line's to say that, I was a prisoner-of-war, I wonder what, the kiddies say to it, and my mother & your mother, Please tell my mum & yours that as soon as I get a chance, I will write to them, & dear old George too, I can quite imagine the surprise on his face, when you told him, Beleive me darling it is taking me a jolly long time to get used to this penned up business, although it can't be helped, I had a good try in getting away when "Tobruk" fell, It took them 9 days after before I was captured and beleive me darling what a thrilling experience, I throughly enjoyed every minute of it. Well Sweetheart I must close now so cheerio & keep smileing

From Your Ever Loveing Husband Arthur

Brian XXX Jeffrey XXX XXXX XXX

No date

Dearest Elsie

Just a few lines darling to say that everything is alright, I am fit and well and hoping that my few lines finds you and the kiddies the same and not forgetting Hazel. Well sweetheart I suppose that you were quite surprised when you got my first few lines to say that I was a prisoner of war. I wonder what the kiddies say to it and my mother and your mother. Please tell my mother and yours that as soon as I get a chance, I will write to them and dear old George too. I can quite imagine the surprise on his face when you told him. Believe me darling it is taking me a jolly long time to get used to the penned-up business, although it can't be helped. I had a good time in getting away when "Tobruk" fell, it took them 9 days after before I was captured and believe me darling what a thrilling experience, I thoroughly enjoyed every minute of it. Well sweetheart I must close now so cheerio and keep smiling.

From your ever-loving Husband Arthur xxxxxxxx

Brian xxx

Jeffrey xxx

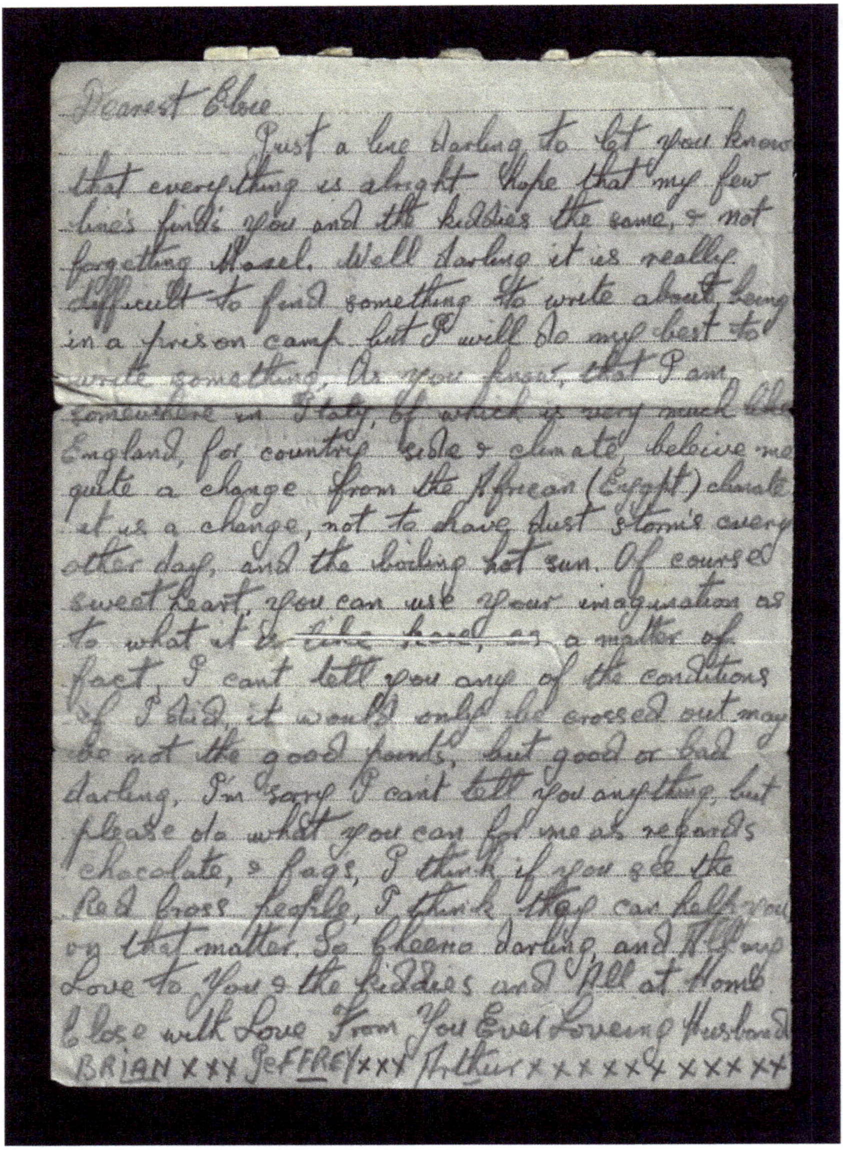

Dearest Elsie

Just a line darling to let you know that everything is alright hope that my few line's find's you and the kiddies the same, & not forgetting Hazel. Well darling it is really difficult to find something to write about being in a prison camp but I will do my best to write something. As you know, that I am somewhere in Italy, of which is very much like England, for country side & climate, beleive me quite a change from the African (Egypt) climate it is a change, not to have dust storm's every other day, and the boiling hot sun. Of course sweet heart, you can use your imagination as to what it is ~~like here~~, as a matter of fact, I can't tell you any of the conditions if I did, it would only be crossed out may be not the good points, but good or bad darling, I'm sorry I can't tell you anything, but please do what you can for me as regards chocolate, & fags, I think if you see the Red Cross people, I think they can help you on that matter. So Cheerio darling, and All my Love to You & the kiddies and All at Home. Close with Love From You Ever Loving Husband

BRIAN xxx JEFFREY xxx Arthur xxxxx4 xxxxx

No Date

Dearest Elsie

Just a line darling to let you know that everything is alright. Hope that my few lines find you and the kiddies the same and not forgetting Hazel. Well darling it is really difficult to find something to write about being in a prison camp, but I will do my best to write something. As you know that I am somewhere in Italy, of which is very much like England, for countryside and climate, believe me quite a change from the Africa (Egypt) climate. It is a change not to have dust storms every other day and the boiling hot sun. Of course sweetheart you can use your imagination as to what it is like here, as a matter of fact I can't tell you any of the conditions if I did it would only be crossed out, may be not the good points, but good or bad darling, I'm sorry I can't tell you anything but please do what you can for me a regards chocolate and fags. I think if you see the Red Cross people I think they can help you on that matter. So, cheerio darling and always love to you and the kiddies and all at home.

Close with love from your ever loving Husband Arthur xxxxxxxxxxx

Brian xxx

Jeffrey xxx

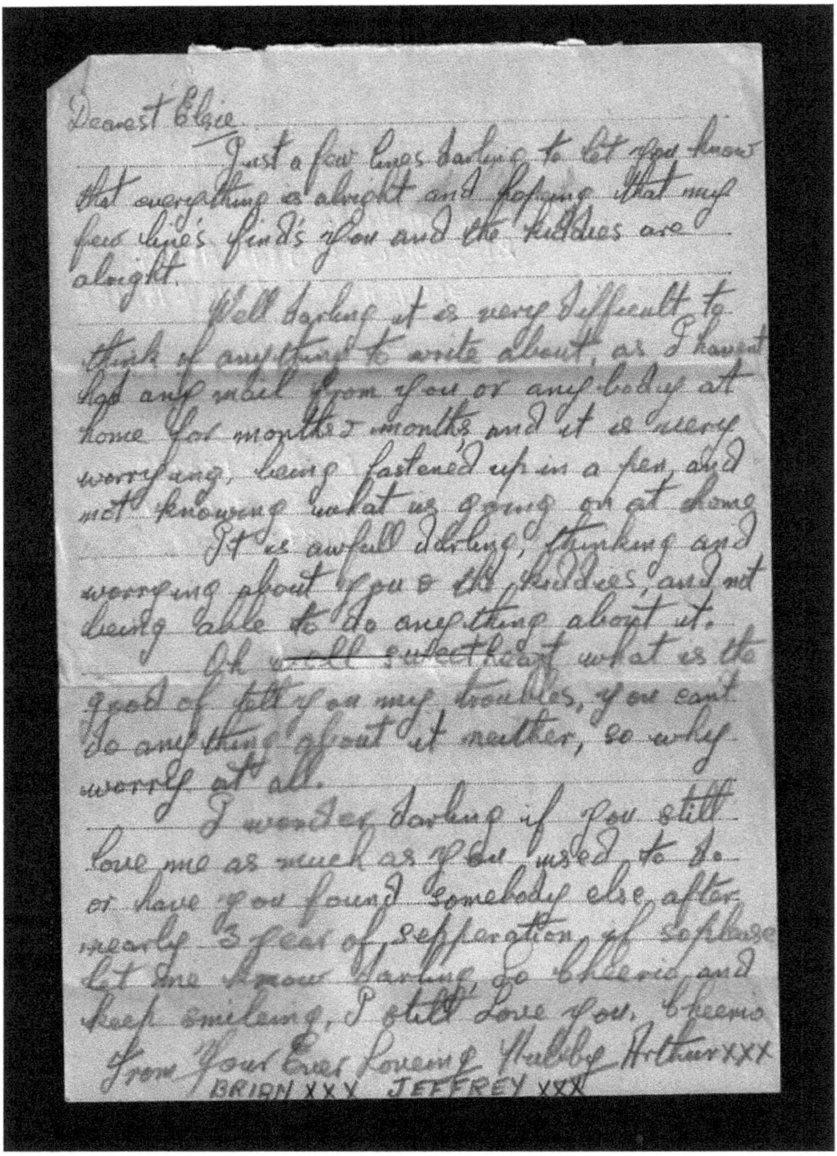

Dearest Elsie

Just a few lines darling to let you know that everything is alright and hoping that my few line's find's you and the kiddies are alright.

Well darling it is very difficult to think of anything to write about, as I haven't had any mail from you or anybody at home for monthes months and it is very worrying, being fastened up in a pen, and not knowing what is going on at home.

It is awfull darling, thinking and worrying about you & the kiddies, and not being able to do anything about it.

Oh ~~well~~ sweetheart what is the good of telling you my troubles, you can't do anything about it neither, so why worry at all.

I wonder darling if you still love me as much as you used to do or have you found somebody else after nearly 3 year of sepperation, if so please let me know darling. So cheerio and keep smileing, I still Love you, Cheerio

from Your Ever Loveing Hubby Arthur XXX

BRIAN XXX JEFFREY XXX

No Date

Dearest Elsie

Just a few lines to let you know that everything is alright and hoping my few lines finds you and the kiddies are alright.

Well darling it is very difficult to think of anything to write about as I haven't had any mail from you or anybody at home for months and month and it is very worrying being fastened up in a pen and not knowing what is going on at home.

It is awful darling, thinking and worrying about you and the kiddies and not being able to do anything about it.

Oh well sweetheart what is the good of telling you my troubles, you can't do anything about it either, so why worry at all.

I wonder darling if you still love me as much as you used to do or have you found somebody else after nearly 3 years of separation if so please let me know darling.

So, cheerio and keep smiling, I still love you cheers from your ever-loving Hubby Arthur xxx

Brian xxx

Jeffrey xxx

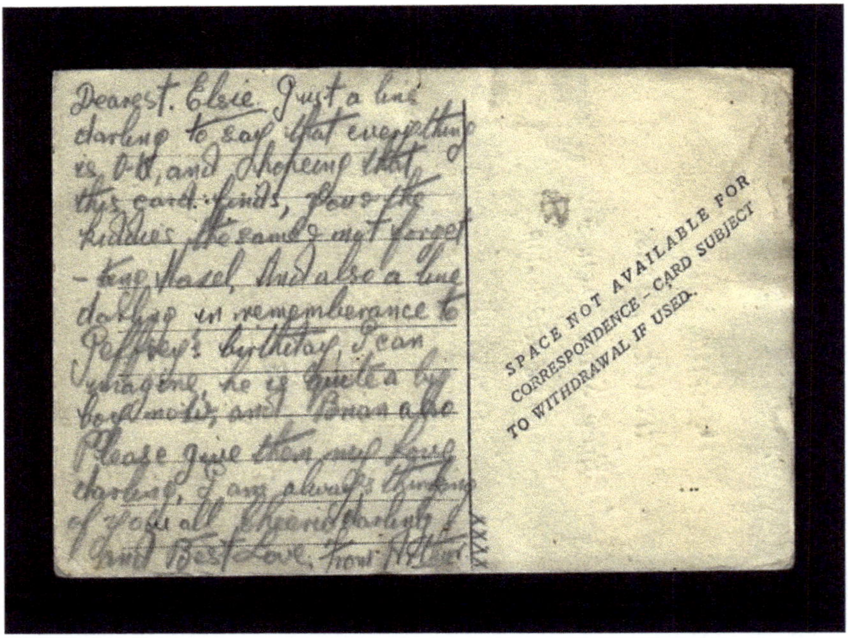

No Date

Dearest Elsie, just a line darling to say that everything is OK and hoping that this card finds you and the kiddies the same and not forgetting Hazel. And also, a line darling in remembrance to Jeffrey's birthday I can imagine he is quite a big boy now and Brian. Please give them my love darling. I am always thinking of you all. Cheerio darling and best love from Arthur xxxx

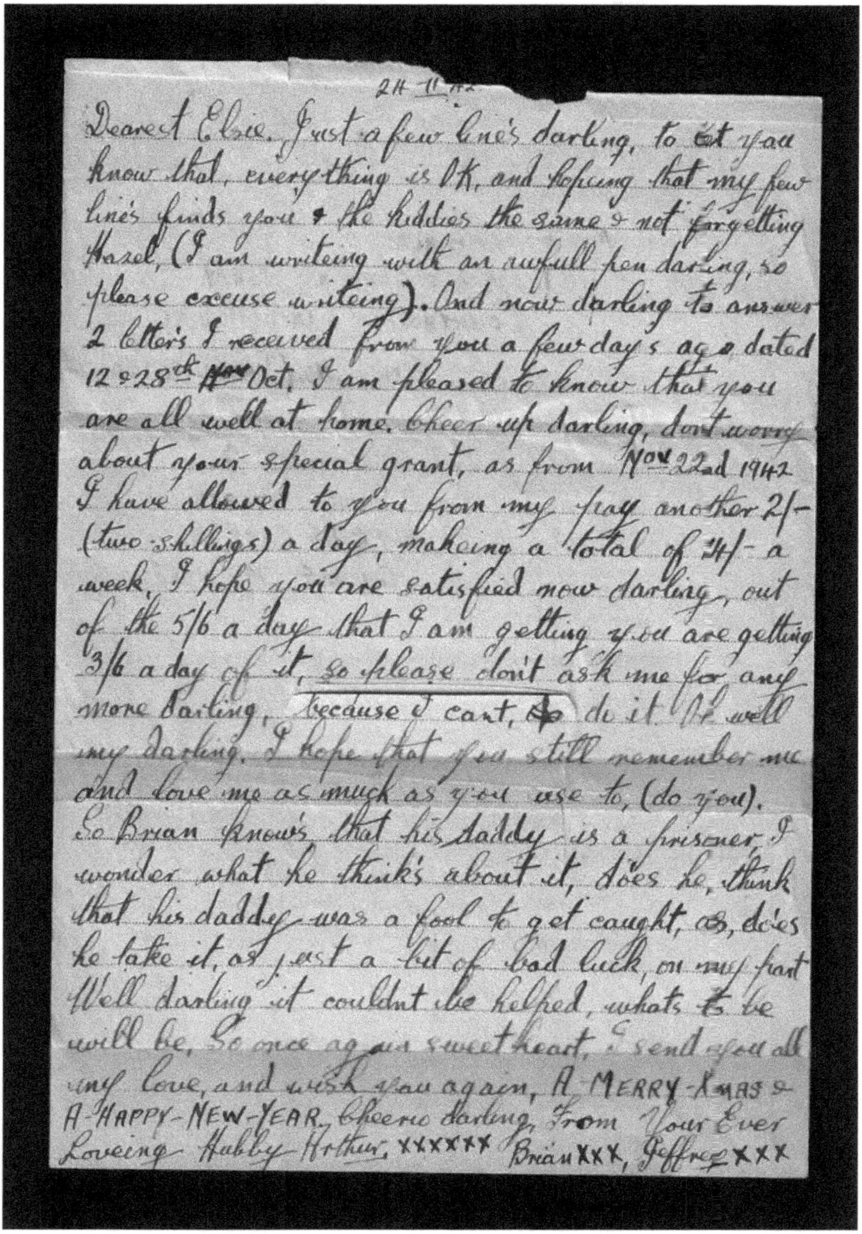

24-11-42

Dearest Elsie. Just a few line's darling, to let you
know that, everything is OK, and hopeing that my few
line's finds you & the kiddies the same & not forgetting
Hazel, (I am writeing with an awfull pen darling, so
please excuse writeing). And now darling to answer
2 letter's I received from you a few day's ago dated
12 & 28th ~~Nov~~ Oct. I am pleased to know that you
are all well at home. Cheer up darling, dont worry
about your special grant, as from Nov 22nd 1942
I have allowed to you from my pay another 2/-
(two shillings) a day, makeing a total of 14/- a
week, I hope you are satisfied now darling, out
of the 5/6 a day that I am getting you are getting
3/6 a day of it, so please don't ask me for any
more darling, because I cant, do it. Well
my darling. I hope that you still remember me
and love me as much as you use to, (do you).
So Brian knows that his daddy is a prisoner, I
wonder what he think's about it, does he, think
that his daddy was a fool to get caught, or, does
he take it, as just a bit of bad luck, on my part
Well darling it couldnt be helped, whats to be
will be, So once again sweetheart, I send you all
my love, and wish you again, A-MERRY-XMAS &
A-HAPPY-NEW-YEAR, Cheerio darling, From Your Ever
Loveing Hubby Arthur, XXXXX Brian XXX, Jeffrey XXX

24th November 1942

Dearest Elsie,

Just a few lines darling to let you know that everything is OK and hoping that my few lines find you and the kiddies the same and not forgetting Hazel. (I am writing with an awful pen darling so please excuse writing). And now darling to answer 2 letters I received from you a few days ago dated 12 & 28th Oct. I am pleased to know that you are all well at home. Cheer up darling don't worry about your special grant as from Nov 22nd, 1942, I have allowed to you from my pay another 2/- (two shillings) a day, making a total of 14/- a week. I hope you are satisfied now darling, out of the 5/6 a day that I am getting you are getting 3/6 a day of it, so please don't ask me for any more darling, because I can't do it Oh well my darling I hope that you still remember me and love me as much as you used to (do you). So, Brian knows that his daddy is a prisoner. I wonder what he thinks about it, does he think that his daddy was a fool to get caught, or does he take it as just a bit of bad luck on my part. Well darling it could not be helped what's to be will be. So once again sweetheart, send you all my love and wish you again A MERRY XMAS and A HAPPY NEW YEAR. Cheerio darling from you ever loving Hubby Arthur xxxxxx

Brian xxx

Jeffrey xxx

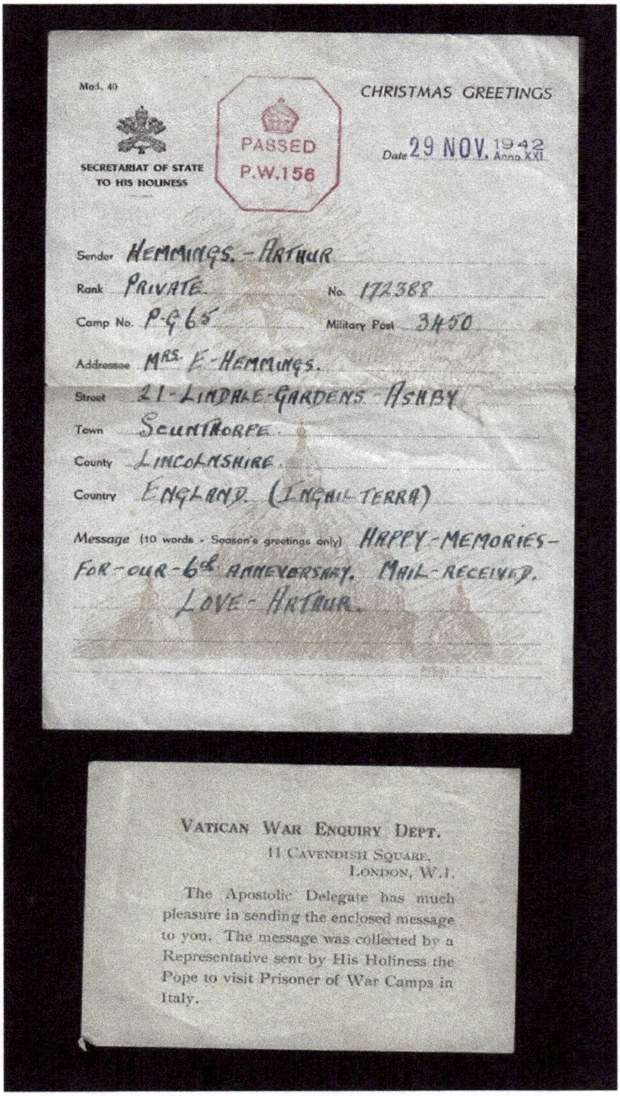

Mod. 40

SECRETARIAT OF STATE
TO HIS HOLINESS

PASSED
P.W.156

CHRISTMAS GREETINGS

Date 29 NOV. 1942 Anno XXI

Sender HEMMINGS. - ARTHUR

Rank PRIVATE. No. 172388

Camp No. P.G 65 Military Post 3450

Addressee MRS. E. HEMMINGS.

Street 21 - LINDALE - GARDENS - ASHBY

Town SCUNTHORPE

County LINCOLNSHIRE

Country ENGLAND. (INGHILTERRA)

Message (10 words - Season's greetings only) HAPPY - MEMORIES - FOR - OUR - 6th ANNEVERSARY. MAIL - RECEIVED. LOVE - ARTHUR.

VATICAN WAR ENQUIRY DEPT.
11 CAVENDISH SQUARE,
LONDON, W.1.

The Apostolic Delegate has much pleasure in sending the enclosed message to you. The message was collected by a Representative sent by His Holiness the Pope to visit Prisoner of War Camps in Italy.

This is a really special letter. The Priests, from the Vatican used to visit the prisoners on special occasions and write these letters. They would be taken to the Vatican and posted to the UK

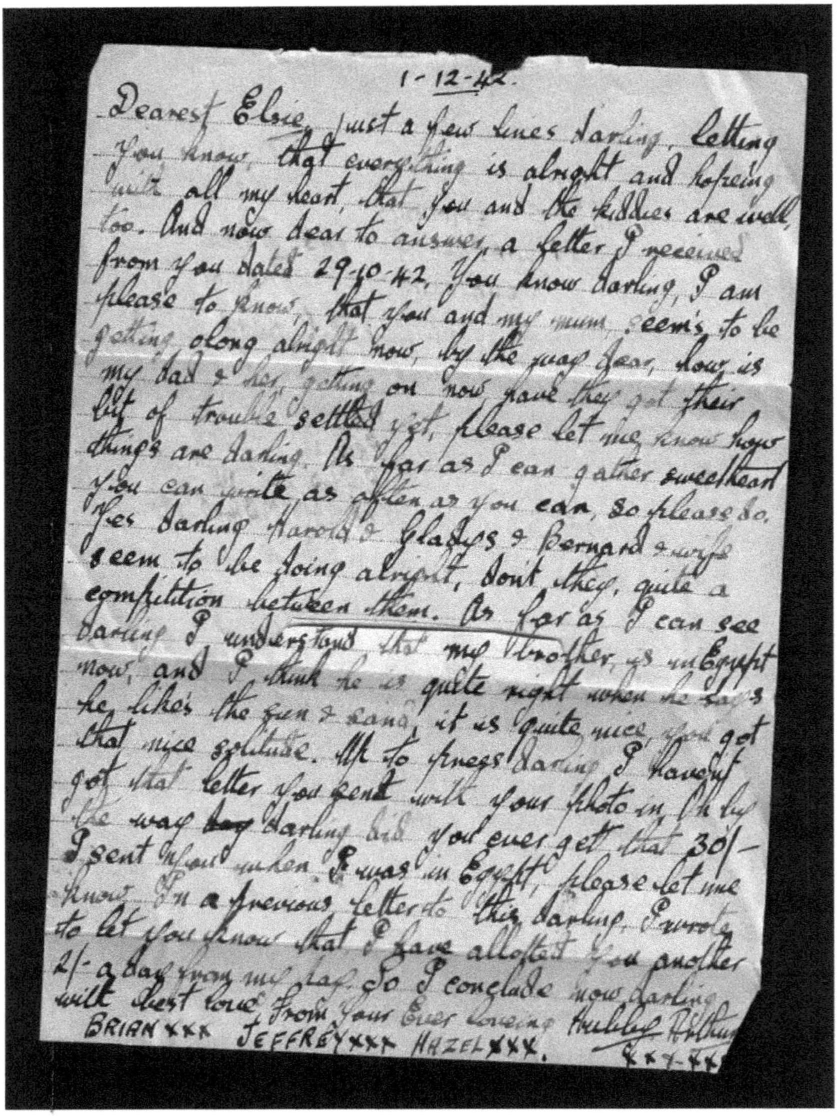

1-12-42.

Dearest Elsie, just a few lines darling, letting
you know, that everything is alright and hopeing
with all my heart, that you and the kiddies are well,
too. And now dear to answer, a letter I received
from you dated 29-10-42, You know darling, I am
please to know, that you and my mum, seem's to be
getting along alright now, by the way dear, how is
my dad & her, getting on now have they got their
bit of trouble settled yet, please let me know how
things are darling. As far as I can gather sweetheart
you can write as often as you can, so please do.
Yes darling Harold & Gladys & Bernard & wife
seem to be doing alright, dont they, quite a
competition between them. As far as I can see
darling I understand that my brother, is in Egypt
now, and I think he is quite right when he says
he like's the sun & sand, it is quite nice, you got
that nice solitude. Up to press darling I haven't
got that letter you sent with your photo in, Oh by
the way ~~boy~~ darling did you ever get that 30/-
I sent you when I was in Egypt, please let me
know. In a previous letter to this darling, I wrote
to let you know that I have alloted you another
2/- a day from my pay. So I conclude now darling
with best love from your Ever loveing Hubby Arthur

BRIAN xxx JEFFREY xxx HAZEL xxx. xxx-xxx

1 -12-42

Dearest Elsie,

Just a few lines darling letting you know that everything is alright and hoping with all my heart that you and the kiddies are well too. And now dear to answer a letter received from you dated 20/10/42. You know darling I am pleased to know that you and your mum seem to be getting along alright now. By the way dear, how is my dad and her getting on now, have they got their bit of trouble settled yet? Please let me know how things are darling. As far as I can gather sweetheart you can write as often as you can so please do. Yes, darling Harold and Gladys and Bernard and wife seem to be doing alright don't they, quite a competition between them. As far as I can see darling, I understand that my brother is in Egypt now and I think he is quite right when he says he likes the sun and sand, it is quite nice you get that nice solitude. Up to press darling I haven't got that letter you send with your photo in. Oh, by the way darling did you get the 30/- I sent you when I was in Egypt? Please let me know. In a previous letter to this, darling, I wrote to let you know I have allotted you another 2/- a day from my pay. So, I conclude now darling with best love from your ever-loving Hubby Arthur xxxxxx

Brian xxx

Jeffrey xxx

Hazel xxx

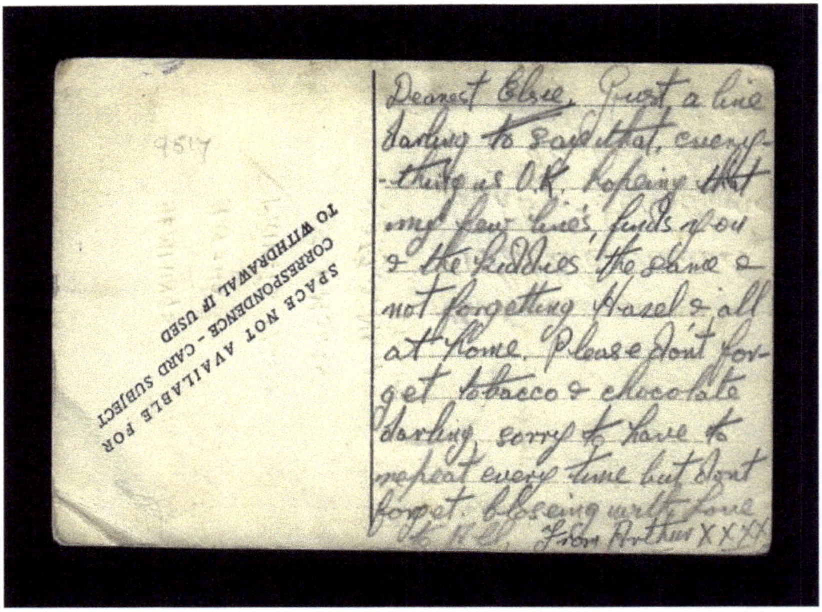

Dearest Elsie, just a line darling to say that everything is OK. Hoping that my few lines find you and the kiddies the same and not forgetting Hazel and all at home. Please don't forget tobacco and chocolate darling, sorry to have to repeat ever time but don't forget. Closing with love to all from Arthur xxxx

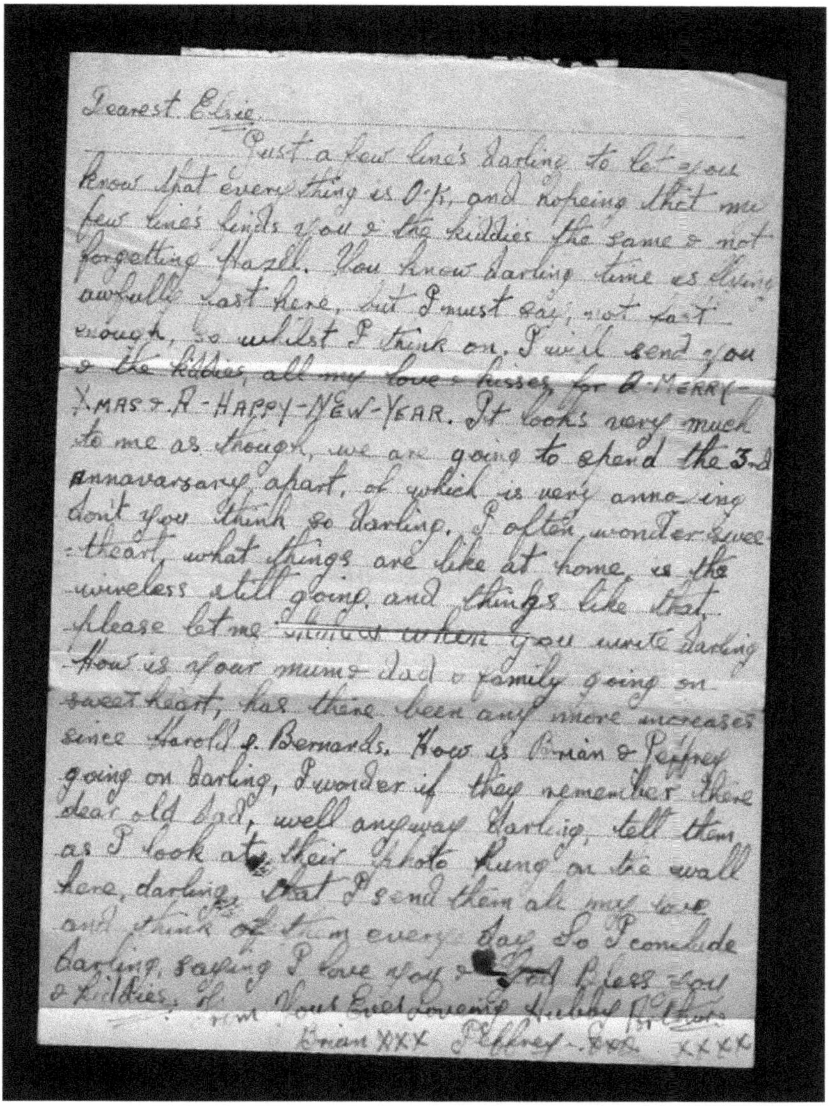

Dearest Elsie

Just a few line's darling to let you know that everything is O.K. and hopeing that me few line's finds you & the kiddies the same & not forgetting Hazel. You know darling time is flying awfully fast here, but I must say, not fast enough, so whilst I think on, I will send you & the kiddies, all my love & kisses, for A-MERRY-XMAS & A-HAPPY-NEW-YEAR. It looks very much to me as though, we are going to spend the 3rd annavarsary apart, of which is very anno-ing don't you think so darling. I often wonder swee-theart, what things are like at home, is the wireless still going, and things like that, please let me ~~know~~ when you write darling How is your mum & dad & family going on sweetheart, has there been any more increases since Harold & Bernards. How is Brian & Jeffrey going on darling, I wonder if they remember there dear old Dad, well anyway darling, tell them as I look at their photo hung on the wall here, darling, that I send them all my love and think of them every day. So I conclude darling, saying I love you & God Bless you & kiddies, from Your Ever Loving Hubby [illegible]

Brian XXX Jeffrey - XXX XXXX

Dearest Elsie

Just a few lines darling to let you know that everything is OK and hoping that my few lines find you and the kiddies the same and not forgetting Hazel. You know darling time is going awful fast here but I must say not fast enough so whilst I think on, I will send you and the kiddies all my love and kisses for A MERRY XMAS AND A HAPPY NEW YEAR. It looks very much to me as though we are going to send the 3rd anniversary apart of which is very annoying don't you think so darling. I often wonder sweetheart what things are like at home is the wireless still going and things like that. Please let me know when you write darling. How is your mum and dad and family going on. Sweetheart has there been any more increases since Harold and Bernards. How is Brian and Jeffrey going on darling, I wonder if they remember their dear old dad. Well anyway darling tell them as I look at their photo hung on the wall here darling, that I send them all my love and think of them every day. So, I conclude darling saying I love you and god bless you and the kiddies. From your ever loving Hubby Arthur. Xxxx

Brian xxx

Jeffrey xxx

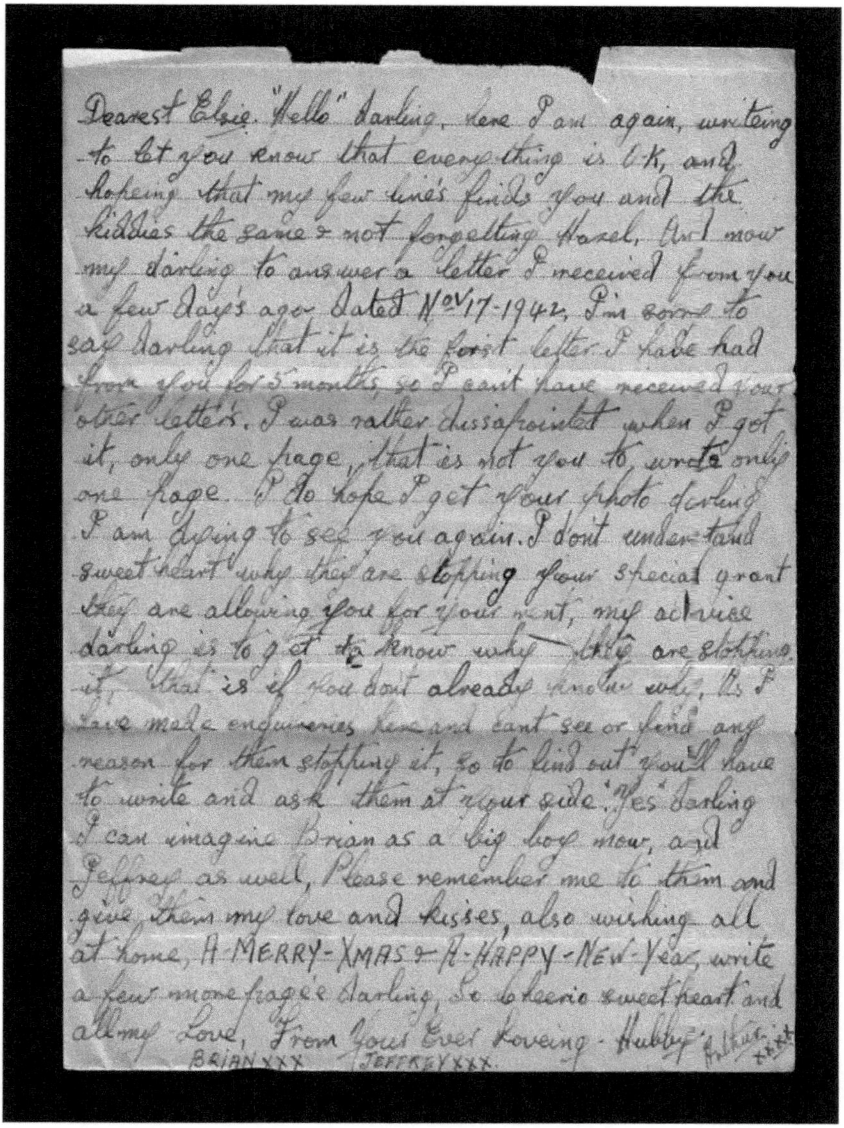

Dearest Elsie. "Hello" darling, here I am again, writing to let you know that everything is O.K, and hopeing that my few line's finds you and the kiddies the same & not forgetting Hazel, And now my darling to answer a letter I received from you a few day's ago dated Nov 17-1942, I'm sorry to say darling that it is the first letter I have had from you for 5 months, so I can't have received your other letter's. I was rather dissapointed when I got it, only one page, that is not you to write only one page. I do hope I get your photo darling I am dying to see you again. I don't under-stand sweet heart why they are stopping your special grant they are allowing you for your rent, my advice darling is to get to know why they are stopping it, that is if you don't already know why, As I have made enquiries here and can't see or find any reason for them stopping it, so to find out you'll have to write and ask them at your side. "Yes" darling I can imagine Brian as a big boy now, and Jeffrey as well, Please remember me to them and give them my love and kisses, also wishing all at home, A-MERRY-XMAS & A-HAPPY-NEW-Year, write a few more page's darling, So Cheerio sweet heart and all my Love, From Your Ever Loveing - Hubby Arthur xxxx

BRIAN XXX JEFFREY XXX.

No date

Dearest Elsie 'Hello' darling here I am again writing to let you know that everything is OK and hoping that my few lines find you and the kiddies the same and not forgetting Hazel. And now my darling to answer a letter I received from you a few days ago dated Nov 17th 1942. I'm sorry to say darling that it is the first letter I have had from you for 5 months, so I cant have received your other letters. I was rather disappointed when I got it only one page that is not you to write only one page. I do hope I get your photo darling I am dying to see you again. I don't understand sweetheart why they are stopping your special grant they are allowing you for your rent, my advice darling is to get to know why they are stopping it, that is if you don't already know why. As I have made enquires here and cant see or find any reason for them stopping it. So to find out you'll have to write and ask them at your side. "Yes" darling I can imagine Brian as a big boy now and Jeffrey as well. Please remember me to them and give them my love and kisses, also wishing all at home A MERRY XMAS & A HAPPY NEW YEAR. Write a few more pages darling. So cheerio sweetheart and all my love from your ever loving Hubby Arthur xxxx

Brian xxx

Jeffrey xxx

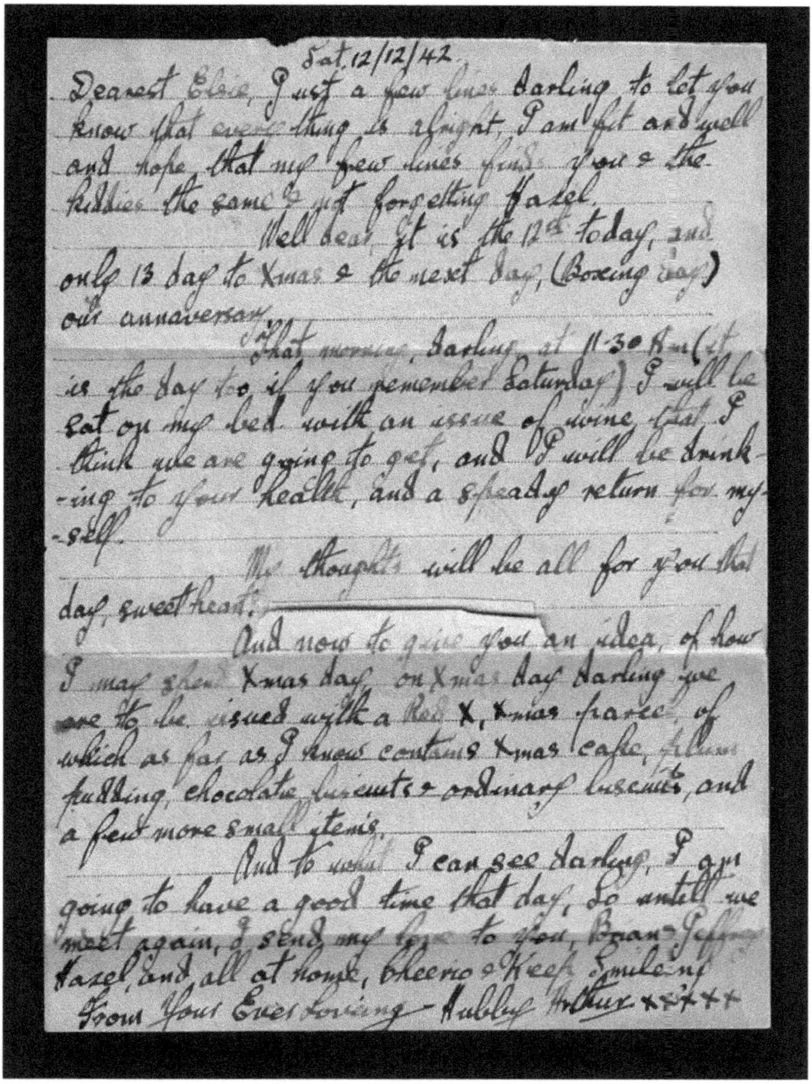

Sat. 12/12/42.

Dearest Elsie, Just a few lines darling to let you know that everything is alright, I am fit and well and hope that my few lines finds you & the kiddies the same & not forgetting Hazel.

Well dear, It is the 12th today, and only 13 day to Xmas & the next day, (Boxing day) our anniversary.

That morning, darling, at 11-30 A.m. (it is the day too, if you remember Saturday) I will be sat on my bed with an issue of wine that I think we are going to get, and I will be drink-ing to your health, and a speedy return for my-self.

My thoughts will be all for you that day, sweetheart.

And now to give you an idea of how I may spend Xmas day, on Xmas day darling we are to be issued with a Red X, Xmas parcel, of which as far as I know contains Xmas cake, plum pudding, chocolate biscuits & ordinary biscuits, and a few more small items.

And to what I can see darling, I am going to have a good time that day, So untill we meet again, I send my love to you, Brian, Jeffrey, Hazel, and all at home, Cheerio & Keep Smileing

From Your Ever Loving - Hubby Arthur xxxxx

Sat 12/12/42

Dearest Elsie, just a few lines darling to let you know that everything is alright. I am fit and well and hope that my few lines find you and the kiddies the same and not forgetting Hazel.

Well dear it is the 12th today and only 13 days to Xmas and the next day (Boxing Day) our anniversary.

That morning darling at 11.30 am (it is the day to if you remember Saturday) I will be sat on my bed with an issue of wine that I think we are going to get and I will be drinking to your health and a speedy return for myself.

My thoughts will be all for you that day sweetheart.

And now to give you an idea of how I may spend Xmas day. On Xmas day darling we are to be issued with a Red X Xmas parcel of which as far as I know contains Xmas cake, plum pudding, chocolate biscuits and ordinary biscuits and a few more small items.

And to until I can see you darling, I am going to have a good time that day. So, until we meet again, I send my love to you, Brian and Jeffrey, Hazel and all at home. Cheerio and keep smiling from your ever-loving Hubby Arthur xxxxx

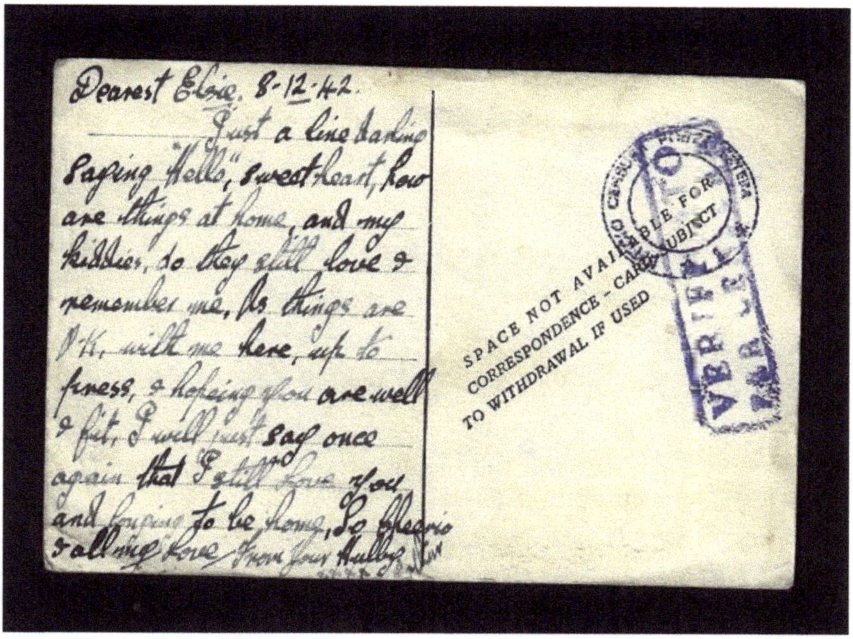
Dearest Elsie. 8-12-42.
Just a line darling
saying "Hello", sweetheart, how
are things at home, and my
kiddies, do they still love &
remember me, As things are
O.K. with me here, up to
press, & hoping you are well
& fit, I will just say once
again that I still love you
and hoping to be home, So cheerio
& all my love from your Hubby Ar

SPACE NOT AVAI
CORRESPONDENCE - CA
TO WITHDRAWAL IF USED

8-12-42

Dearest Elsie

Just a line darling saying "Hello" sweetheart how are things at home do they still love and remember me. As things are OK with me here up to press and hoping you are well and fit. I will just say once again that I still love you and hoping to be home. So cheerio and all my love from your Hubby Arthur xxxx

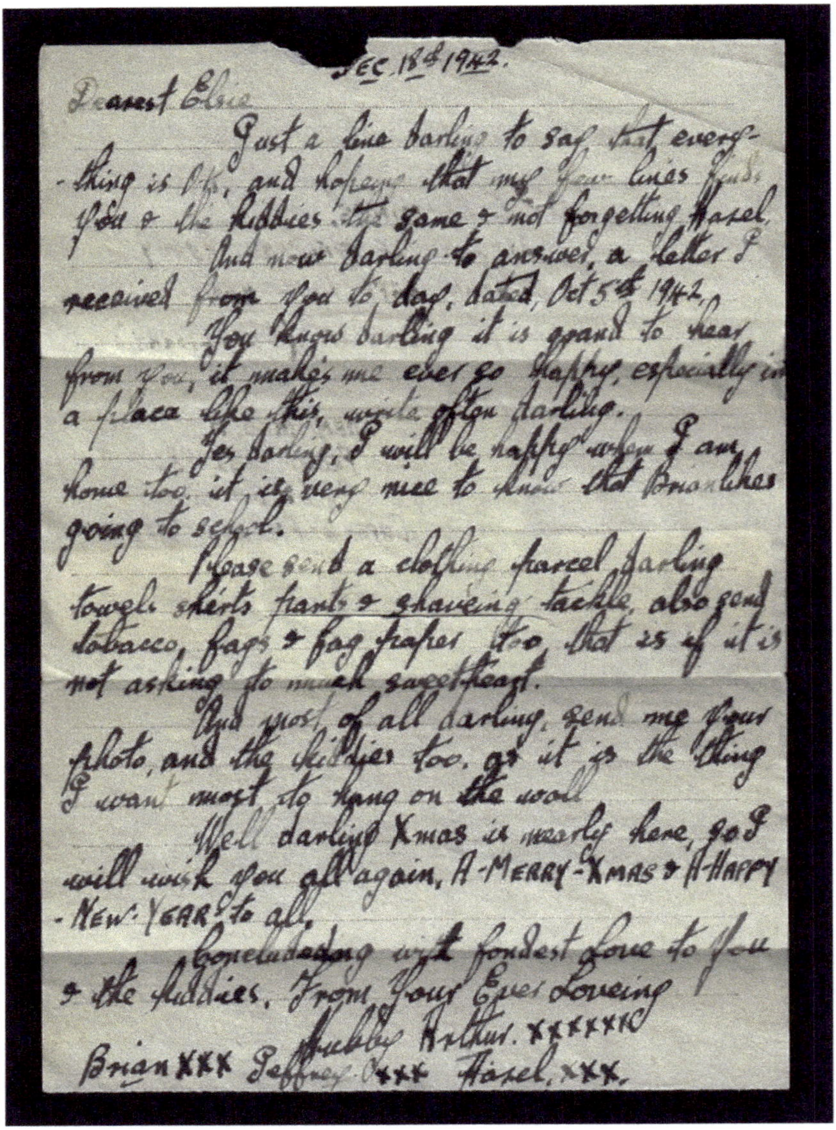

Dec. 18th 1942.

Dearest Elsie

Just a line darling to say that every-thing is O.K. and hoping that my few lines finds you & the kiddies the same & not forgetting Hazel.

And now darling to answer a letter I received from you to day, dated, Oct 5th 1942,

You know darling it is grand to hear from you, it makes me ever so happy, especially in a place like this, write often darling.

Yes darling, I will be happy when I am home too, it is very nice to know that Brian likes going to school.

Please send a clothing parcel darling towels shirts pants & shaveing tackle, also send tobacco, fags & fag paper too, that is if it is not asking to much sweetheart.

And most of all darling, send me your photo, and the kiddies too, as it is the thing I want most to hang on the wall

Well darling Xmas is nearly here, so I will wish you all again, A-MERRY-XMAS & A-HAPPY-NEW-YEAR to all,

Concludeding with fondest love to you & the kiddies. From Your Ever Loveing

Hubby Arthur. XXXXXX

Brian XXX Jeffrey XXX Hazel. XXX.

Dec 18th 1942

Dearest Elsie,

Just a line darling to say that everything is OK and hoping that my few lines find you and the kiddies the same and not forgetting Hazel.

And now darling to answer a letter I received from you today dated Oct 5th, 1942.

You know darling it is good to hear from you; it makes me ever so happy especially in a place like this, write often darling.

Yes, darling I will be happy when I am home too. It is very nice to know that Brian likes going to school.

Please send a clothing parcel darling, towels, shirts, pants and shaving tackle also send tobacco, fags and fag paper too. That is if it is not asking too much sweetheart.

And most of all darling send me your photo and the kiddies too as it is the thing I want most to hand on the wall.

Well darling Xmas is nearly here so I will wish you all again A MERRY XMAS & A HAPPY NEW YEAR to all.

Concluding with fondest love to you and the kiddies from your ever-loving Hubby Arthur xxxxxx

Brian xxx

Jeffrey xxx

Hazel xxx

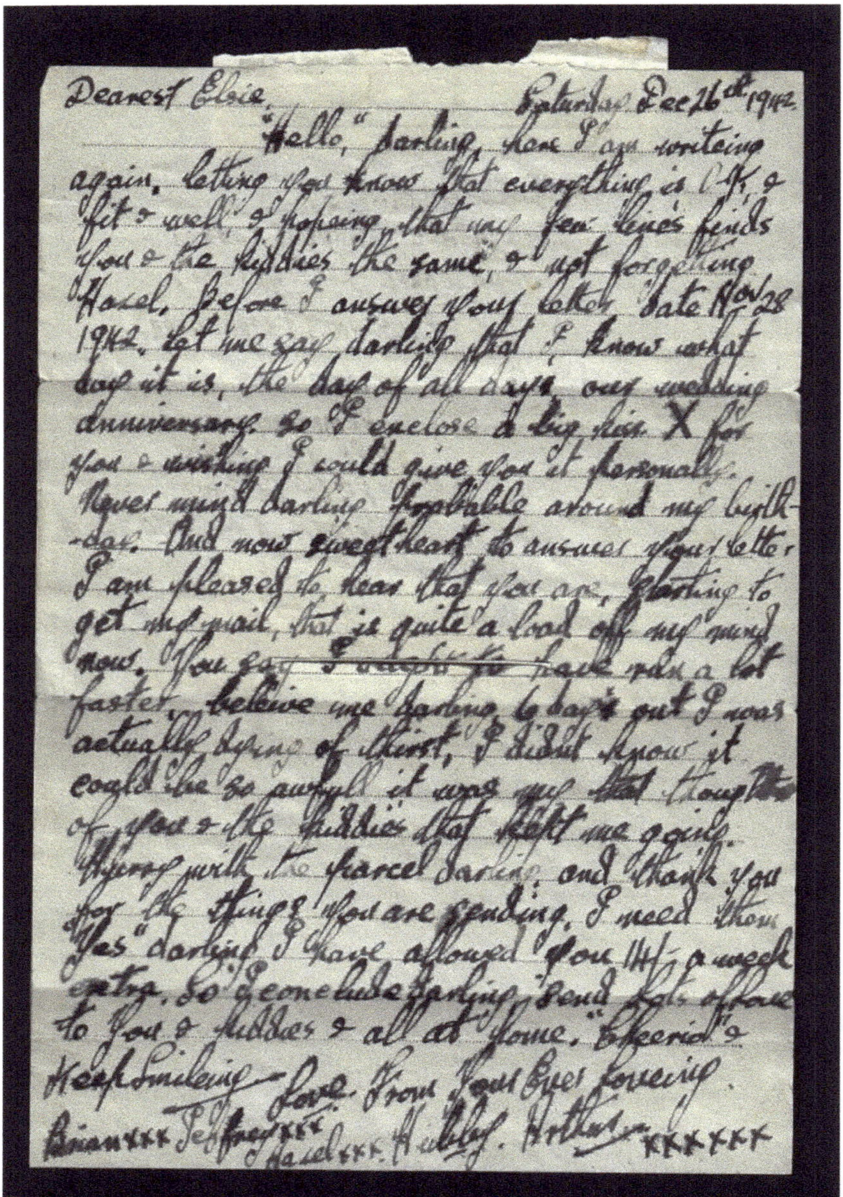

Dearest Elsie, Saturday Dec 26th 1942.

"Hello" darling, here I am writeing again, letting you know that everything is O.K. & fit & well, & hopeing that my few lines finds you & the kiddies the same, & not forgetting Hazel, before I answer your letter date Nov 28 1942. let me say darling that I know what day it is, the day of all days, our wedding anniversary. so I enclose a big kiss X for you & wishing I could give you it personally. Never mind darling, probable around my birth-day. And now sweetheart to answer your letter. I am pleased to hear that you are starting to get my mail, that is quite a load off my mind now. You say I [illegible] to have ran a lot faster, believe me darling 6 day's out I was actually dying of thirst, I didn't know it could be so awfull it was my thoughts of you & the kiddies that kept me going. Sorry with the parcel darling, and thank you for the things you are sending. I need them "Yes" darling I have allowed you 14/- a week extra. So I conclude darling. Send lots of love to you & kiddies & all at Home. "Cheerio" & Keep Smileing

Love. From Your Ever Loveing. Hubby. Arthur XXXXXX

Brian XXX Jeffrey XXX Hazel XXX

Saturday Dec 26th, 1942

Dearest Elsie

"Hello" darling here I am writing again letting you know that everything is OK and fit and well and hoping that my few lines find you and the kiddies the same and not forgetting Hazel. Before I answer your letter dated Nov 28, 1942, let me say darling that I know what day it is the day of all days our wedding anniversary so I enclose a big kiss X for you and wishing I could give you it personally. Never mind darling probably around my birthday. And now sweetheart to answer your letter I am pleased to hear that you are starting to get my mail that is quite a load off my mind now. You say I ought to have ran a lot faster, believe me darling 6 days out I was actually dying of thirst. I didn't know it could be so awful it was my thought of you and the kiddies that kept me going. Hurry with the parcel darling and thank you for the things you are sending. I need them "yes" darling I have allowed you 14/- a week extra. So, I conclude darling send lots of love to you and the kiddies and all at home. Cheerio and keep smiling love from your ever-loving Hubby Arthur xxxxxx

Brian xxx

Jeffrey xxx

Hazel xxx

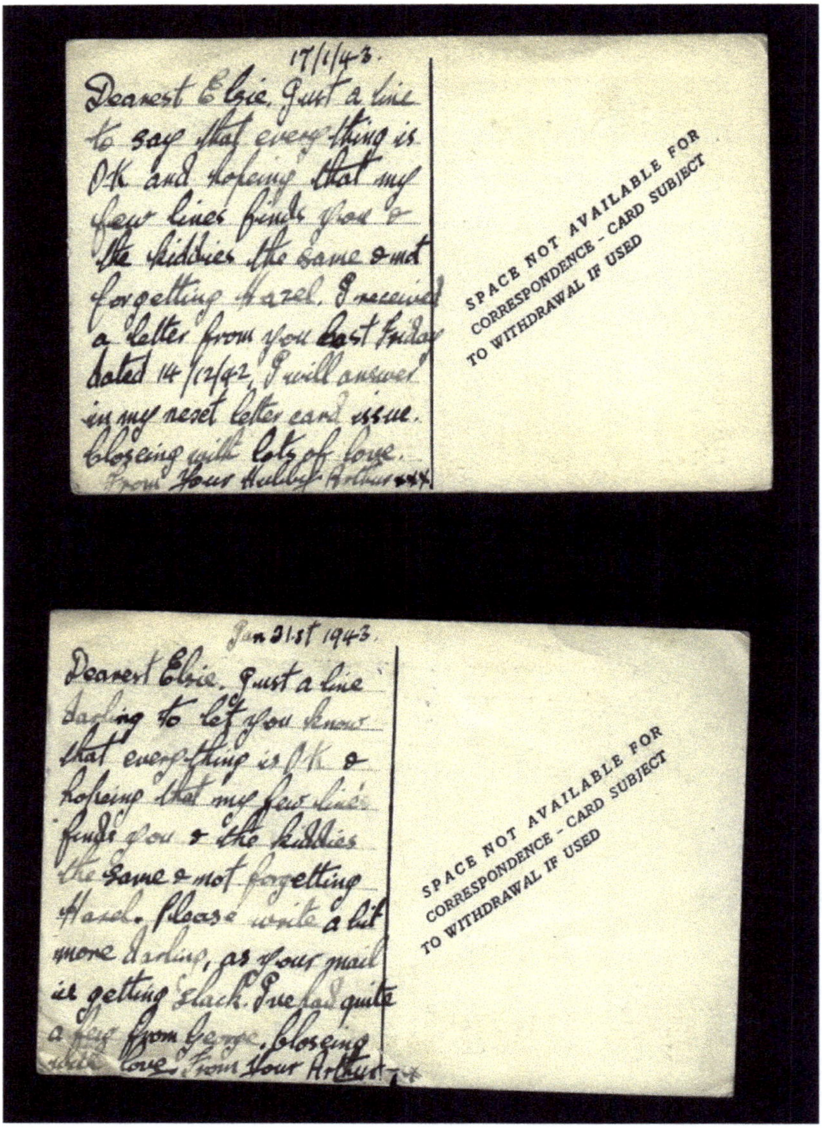

17/1/43.
Dearest Elsie, Just a line
to say that everything is
OK and hopeing that my
few lines finds you &
the kiddies the same & not
forgetting Hazel. I received
a letter from you last Friday
dated 14/12/42, I will answer
in my next letter card issue.
Closeing with lots of love.
From Your Hubby Arthur xxx

SPACE NOT AVAILABLE FOR
CORRESPONDENCE - CARD SUBJECT
TO WITHDRAWAL IF USED

Jan 31st 1943.
Dearest Elsie, Just a line
darling to let you know
that everything is OK &
hopeing that my few lines
finds you & the kiddies
the same & not forgetting
Hazel. Please write a bit
more darling, as your mail
is getting slack. I've had quite
a few from George. Closeing
with love from your Arthur xx

SPACE NOT AVAILABLE FOR
CORRESPONDENCE - CARD SUBJECT
TO WITHDRAWAL IF USED

17/1/43

Dearest Elsie. Just a line to say that everything is OK and hoping that my few lines find you and the kiddies the same and not forgetting Hazel. I received a letter from you last Friday dated 14/12/42. I will answer it in my next letter card issue. Closing with lots of love from your Hubby Arthur xxx

Jan 21st, 1943

Dearest Elsie. Just a line darling to let you know that everything is OK and hoping that my few lines find you and the kiddies the same and not forgetting Hazel. Please write a bit more darling as your mail is getting stuck. I've had quite a few from George. Closing with love from your Arthur xx

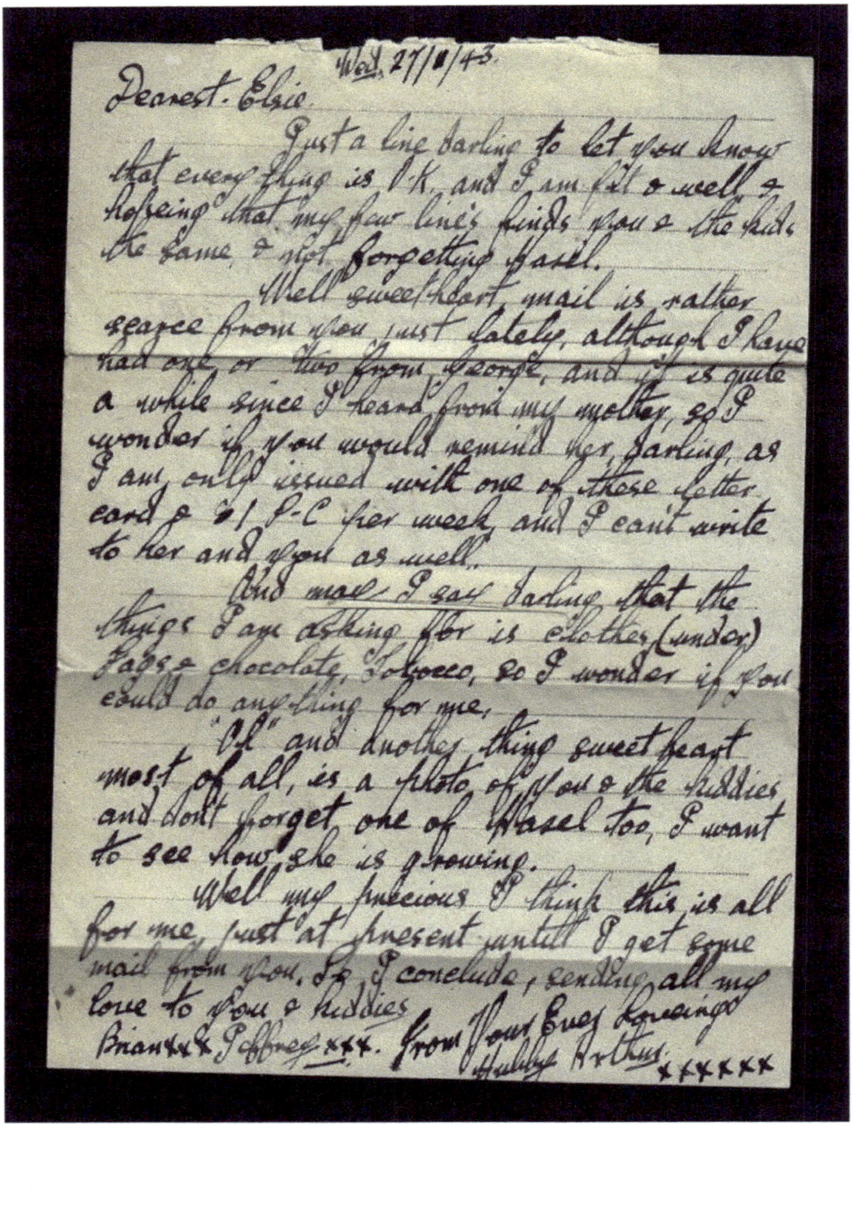

Wed. 27/11/43.

Dearest. Elsie

Just a line darling to let you know that every thing is O.K. and I am fit & well & hopeing that my few line's finds you & the kids the same, & not forgetting Hazel.

Well sweetheart, mail is rather scarce from you just lately, although I have had one or two from George, and it is quite a while since I heard from my mother, so I wonder if you would remind her, darling, as I am only issued with one of these letter card & 1 P-C per week, and I can't write to her and you as well.

And may I say darling that the things I am asking for is clothes (under) fags & chocolate, tobacco, so I wonder if you could do any-thing for me.

"Oh" and another thing sweet heart most of all, is a photo, of you & the kiddies and don't forget one of Hazel too, I want to see how she is growing.

Well my precious I think this is all for me, just at present until I get some mail from you. So I conclude, sending all my love to you & kiddies

Brian xxx Geoffrey xxx. From Your Ever Loving Hubby Arthur. xxxxxx

Wed 27/1/43

Dearest Elsie

Just a line darling to let you know that everything is OK and I am fit and well and hoping that my few lines find you and the kids the same and not forgetting Hazel.

Well sweetheart mail is rather sparce from you just lately although I have had one or two from George and it is quite a while since I heard from my mother so I wonder if you would remind her darling as I am only issued with one of these letter cards 1PC per week and I can't write to her and you as well.

And may I say darling that the things I am asking for is clothes (under) fags and chocolate, tobacco so I wonder if you could do anything for me.

Oh, and another thing sweetheart most of all is a photo of you and the kiddies and don't forget one of Hazel too. I want to see how she is growing.

Well, my precious I think this is all for me just at present until I get some mail from you. So, I conclude sending all my love to you and kiddies from your ever loving Hubby Arthur xxxxxx

Brian xxx

Jeffrey xxx

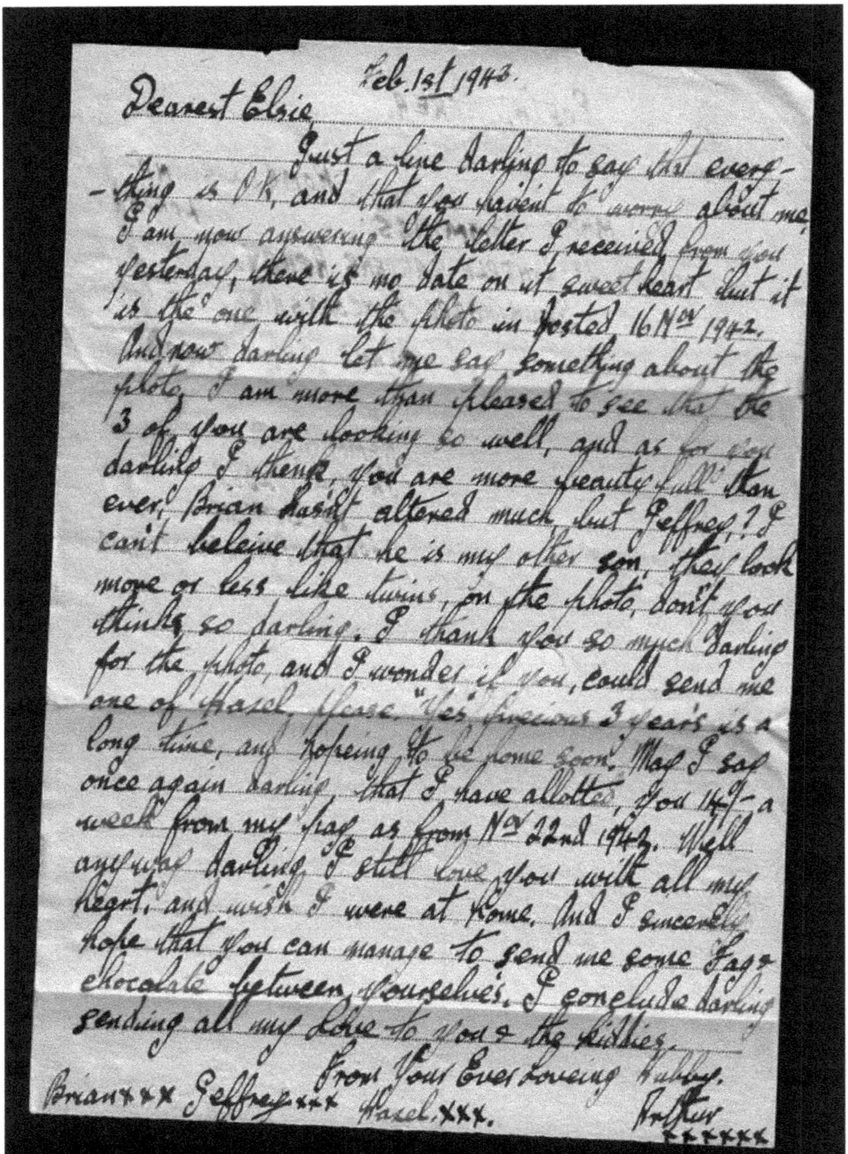

Feb. 1st 1943.

Dearest Elsie,

Just a line darling to say that every-
-thing is O.K., and that you haven't to worry about me,
I am now answering the letter I received from you
yesterday, there is no date on it sweetheart but it
is the one with the photo in posted 16 Nov 1942.
And now darling let me say something about the
photo, I am more than pleased to see that the
3 of you are looking so well, and as for you
darling I think, you are more beauty full than
ever, Brian hasn't altered much, but Geffrey? I
can't beleive that he is my other son, they look
more or less like twins, on the photo, don't you
think so darling. I thank you so much darling
for the photo, and I wonder if you, could send me
one of Hazel please. "Yes" precious 3 years is a
long time, and hopeing to be home soon. May I say
once again darling, that I have alloted you 14/- a
week from my pay as from Nov 22nd 1942. Well
anyway darling I still love you with all my
heart, and wish I were at home. And I sincerely
hope that you can manage to send me some fags &
chocolate between yourselves. I conclude darling
sending all my love to you & the kiddies.

From Your Ever Loveing Hubby.

Brian xxx Geffrey xxx Hazel. xxx. Walter xxxxxx

Feb 1st, 1943

Dearest Elsie,

Just a line darling to say that everything is OK and that you haven't to worry about me I am now answering the letter I received from you yesterday. There is no date on it sweetheart, but it is the one with the photo in, posted 16th Nov 1942. And now darling let me say something about the photo I am more than pleased to see that you 3 are looking so well and as for you darling I think you are more beautiful than ever. Brian hasn't altered much but Jeffrey? I can't believe that he is my other son, they both look more or less like twins, on the photo don't you thinks so darling? I thank you so much darling for the photo and I wonder if you could send me one of Hazel please. "Yes" precious 3 years is a long time and hoping to be home soon. May I say once again darling that I have allotted you 14/- a week from my pay as from Nov 22nd 1942. Well anyway darling I still love you with all my heart and wish I were at home and I sincerely hope that you can manage to send me some fags and chocolate between yourselves. I conclude darling sending all my love to you and the kiddies from your ever-loving Hubby Arthur xxxxxx

Brian xxx

Jeffrey xxx

Hazel xxx

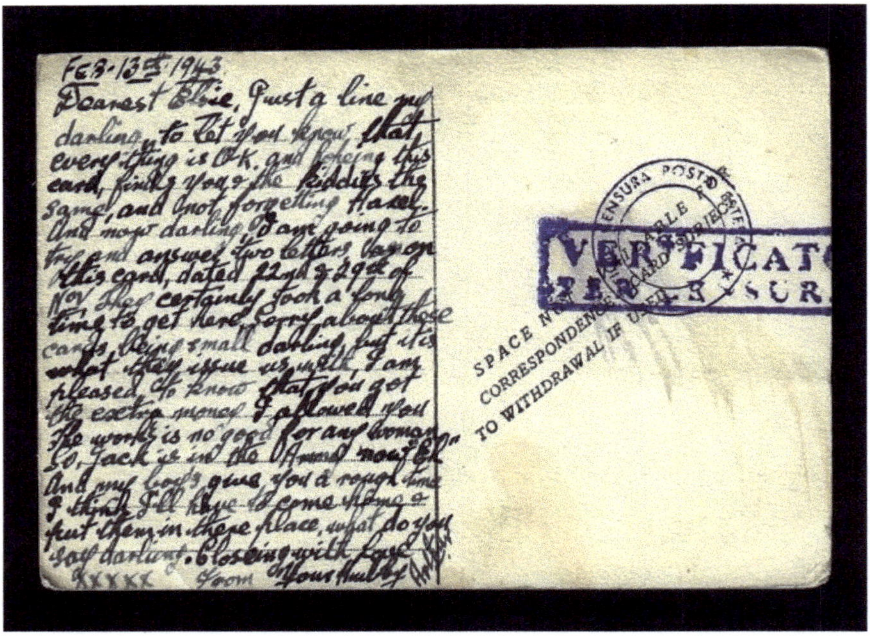

Feb 13th, 1943

Dearest Elsie, just a line my darling to let you know that everything is OK and hoping this card finds you and the kiddies the same and not forgetting Hazel. And now darling I am going to try to answer two letters on this card dated 22nd and 29th of Nov. They certainly took a long time to get here. Sorry about these cards being small darling but it's what they issue us with. I am pleased to know that you got the extra money I allowed you. The works is no good for any woman. So, Jack is in the Army now "ek". And my boys give you a rough time I think I'll have to come home and put them in their place, what do you say darling. Closing with love your Hubby Arthur xxxxx

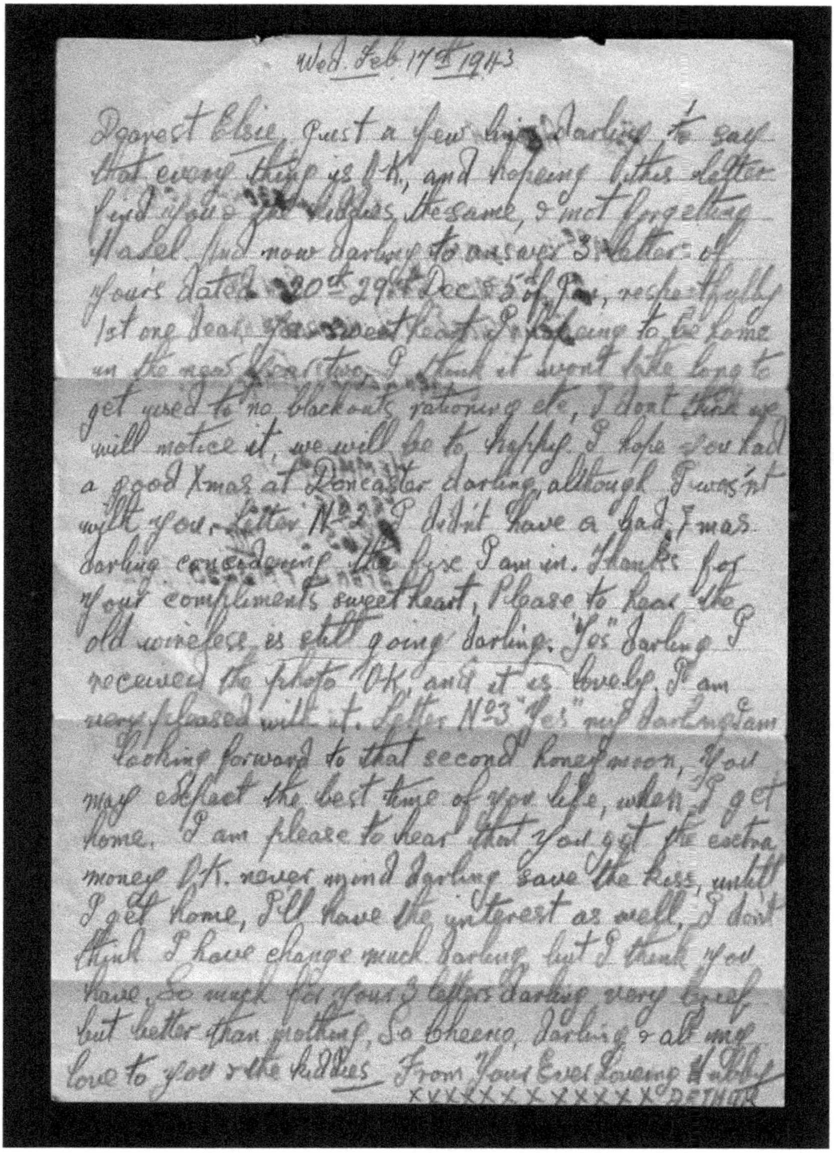

Wed. Feb 17th 1943

Dearest Elsie, just a few lines darling to say that every thing is O.K., and hopeing this letter finds you & the kiddies the same, & not forgetting Hazel. And now darling to answer 3 letters of your's dated 20th 29th Dec & 5th of Jan, respectfully 1st one dear, yes sweetheart I'm longing to be home in the next year or two. I think it won't take long to get used to no blackouts rationing etc, I don't think we will notice it, we will be to happy. I hope you had a good Xmas at Doncaster darling although I was'nt with you. Letter No 2 I didn't have a bad Xmas darling considering the fix I am in. Thanks for your compliments sweetheart, Please to hear the old wireless is still going darling. "Yes" darling I received the photo O.K. and it is lovely. I am very pleased with it. Letter No 3 "Yes" my darling I am looking forward to that second honeymoon, You may expect the best time of you life, when I get home. I am please to hear that you got the extra money O.K. never mind darling save the kiss, until I get home, I'll have the interest as well, I don't think I have change much darling, but I think you have, So much for your 3 letters darling very brief but better than nothing, So Cheerio, Darling & all my love to you & the kiddies From Your Ever Loveing Hubby

XXXXXXXXXXXX

Wed Feb 17th, 1943

Dearest Elsie. Just a few lines darling to say that everything is OK and hoping this letter find you and the kiddies the same and not forgetting Hazel. And now darling to answer 3 letters of yours dated 20th, 29th Dec and 5 of Jan, respectfully 1st one dear, yes sweetheart I'm hoping to be home in the new year too. I think it wont take long to get used to no blackouts, rationing, etc. I don't think we will notice it we will be too happy. I hope you had a good Xmas at Doncaster darling although I wasn't with you. Letter No 2 I didn't have a bad Xmas darling considering the situation I am in. Thanks for your compliment's sweetheart. Please to hear the old wireless is still going darling. "yes" darling I received the photo OK, and it is lovely. I am very pleased with it. Letter No 3 "Yes" my darling I am looking forward to that second honeymoon; you may expect the best time of your life when I get home. I am pleased to hear that you're getting the extra money OK. Never mind darling save the kiss until I get home. I'll have the interest as well. I don't think I have changed much darling, but I think you have. So much for your 3 letters darling very brief but better than nothing. So, cheerio darling and all my love to you and the kiddies from your ever loving Hubby Arthur xxxxxxxxxxxx

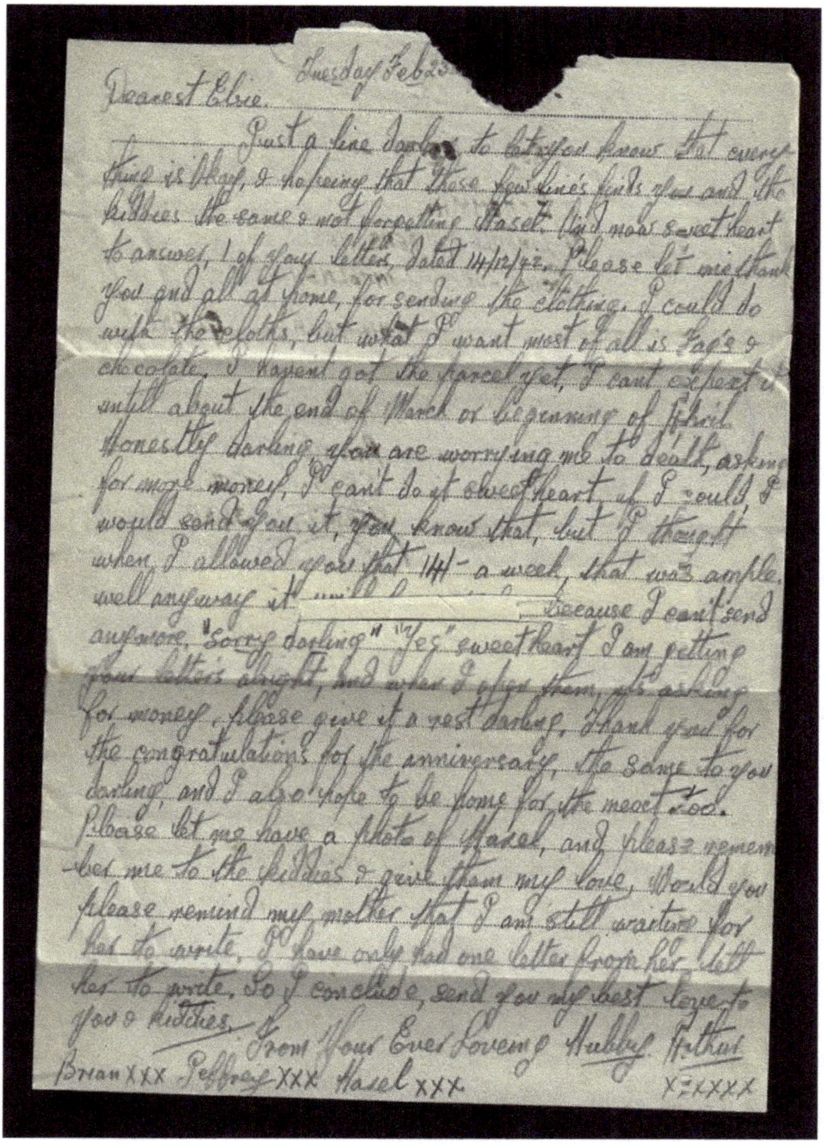

Tuesday Feb 23

Dearest Elsie.

Just a line darling to let you know that everything is okay, & hopeing that these few line's finds you and the kiddies the same & not forgetting Hazel. Had now sweetheart to answer, 1 of your letter's dated 14/12/42. Please let me thank you and all at home, for sending the clothing. I could do with the cloths, but what I want most of all is fag's & chocolate. I havent got the parcel yet, I cant expect it untill about the end of March or beginning of April. Honestly darling you are worrying me to death, asking for more money, I cant do it sweetheart, if I could I would send you it, you know that, but I thought when I allowed you that 14/- a week, that was ample. well anyway it [illegible] because I can't send anymore. "Sorry darling" "Yes" sweetheart I am getting your letter's alright, but when I open them, its asking for money, please give it a rest darling, Thank you for the congratulation's for the anniversary, the same to you darling, and I also hope to be home for the next too. Please let me have a photo of Hazel, and please remember me to the kiddies & give them my love, Would you please remind my mother that I am still waiting for her to write, I have only had one letter from her, tell her to write, So I conclude, send you my best love to you & kiddies,

From Your Ever Loving Hubby Arthur

Brian XXX Jeffrey XXX Hazel XXX

X=XXXX

Tuesday Feb 23rd, 1943

Dearest Elsie

Just a line darling to let you know that everything is okay and hoping that these few lines find you and the kiddies the same and not forgetting Hazel. And now sweetheart to answer 1 of your letters dated 14/12/42. Please let me thank you and all at home for sending the clothing, I could do with the clothes, but what I want most of all is fags and chocolate. I haven't got the parcel yet. I can't expect it until about the end of March or beginning of April. Honestly darling you worrying me to death asking for more money. I can't do it sweetheart if I could I would send you it you know that, but I thought when I allowed you that 14/- a week that was ample. Well anyway it will have to do because I can't send anymore "sorry darling". 'Yes" sweetheart I am getting your letters alright and when I open them its asking for money. Please give it a rest darling. Thank you for the congratulations for the anniversary, the same to you darling and I also hope to be home for the next, too. Please let me have a photo of Hazel and please remember me to the kiddies and give them my love. Would you please remind my mother that I'm still waiting for her to write. I have only had one letter from her, tell her to write. So, I conclude, send you my best love to you and the kiddies.

From your ever-loving Hubby Arthur xxxxxx

Brian xxx

Jeffrey xxx

Hazel xxx

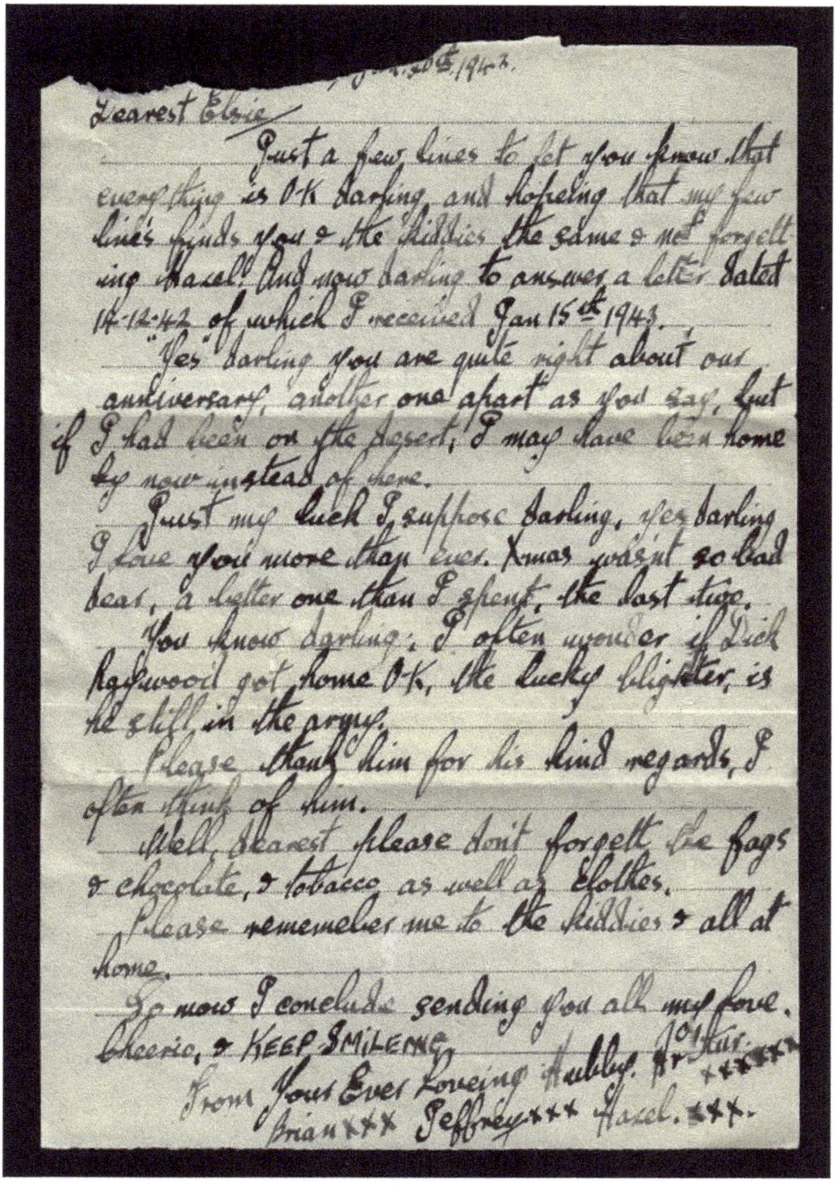

[illegible] 20th, 1943.

Dearest Elsie

Just a few lines to let you know that everything is O.K darling, and hopeing that my few line's finds you & the kiddies the same & not forgetting Hazel. And now darling to answer a letter dated 14-12-42 of which I received Jan 15th 1943.

"Yes" darling you are quite right about our anniversary, another one apart as you say, but if I had been on the desert, I may have been home by now instead of here.

Just my luck I suppose darling, yes darling I love you more than ever. Xmas wasn't so bad dear, a better one than I spent the last time.

You know darling, I often wonder if Dick Heywood got home O.K, the lucky blighter, is he still in the army.

Please thank him for his kind regards, I often think of him.

Well dearest please don't forget the fags & chocolate, & tobacco as well as clothes.

Please remember me to the kiddies & all at home.

So now I conclude sending you all my love. Cheerio, & KEEP SMILEING

From Your Ever Loveing Hubby. Arthur. xxxxx

Brian xxx Geffrey xxx Hazel. xxx.

1943

Dearest Elsie

Just a few lines to let you know that everything is OK darling and hoping that my few lines find you and the kiddies the same and not forgetting Hazel. And now darling to answer a letter dated 14.12.1942 of which I received Jan 15th, 1943.

"Yes" darling you are quite right about our anniversary another one apart as you say but if I had been on the desert, I may have been home by now instead of here.

Just my luck I suppose darling, yes darling I love you more than ever. Xmas wasn't so bad dear, a better one than I spent the last too.

You know darling I often wonder if Dick Haywood got home OK, the lucky blighter, is he still in the army.

Please thank him for his kind regards I often think if him.

Well dearest please don't forget the fags and chocolate and tobacco as well as clothes.

Please remember me to the kiddies and all at home.

So now I conclude sending you all my love. Cheerio and KEEP SMILING.
From your ever-loving Hubby Arthur xxxxxx

Brian xxx

Jeffrey xxx

Hazel xxx

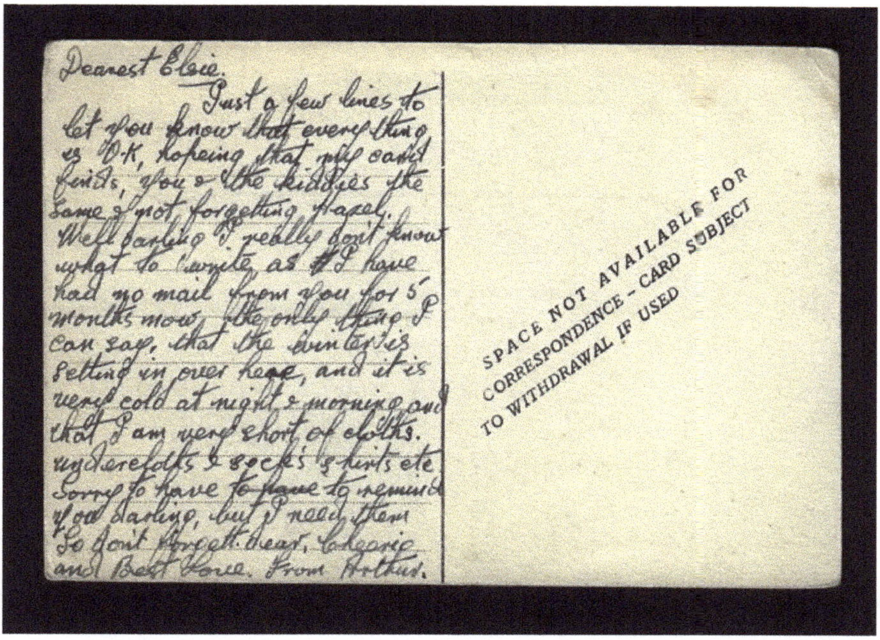

Dearest Elsie,

Just a few lines to let you know that everything is OK, hoping that my card finds you and the kiddies the same and not forgetting Hazel. Well darling I really don't know what to write as I have had no mail from you for 5 months now, the only thing I can say that the winter is setting in over here and it is very cold at night and morning and that I am very short of cloths, underwear and socks, shirts etc. Sorry to have to remind you darling, but I need them so don't forget dear. Cheerio and best love from Arthur

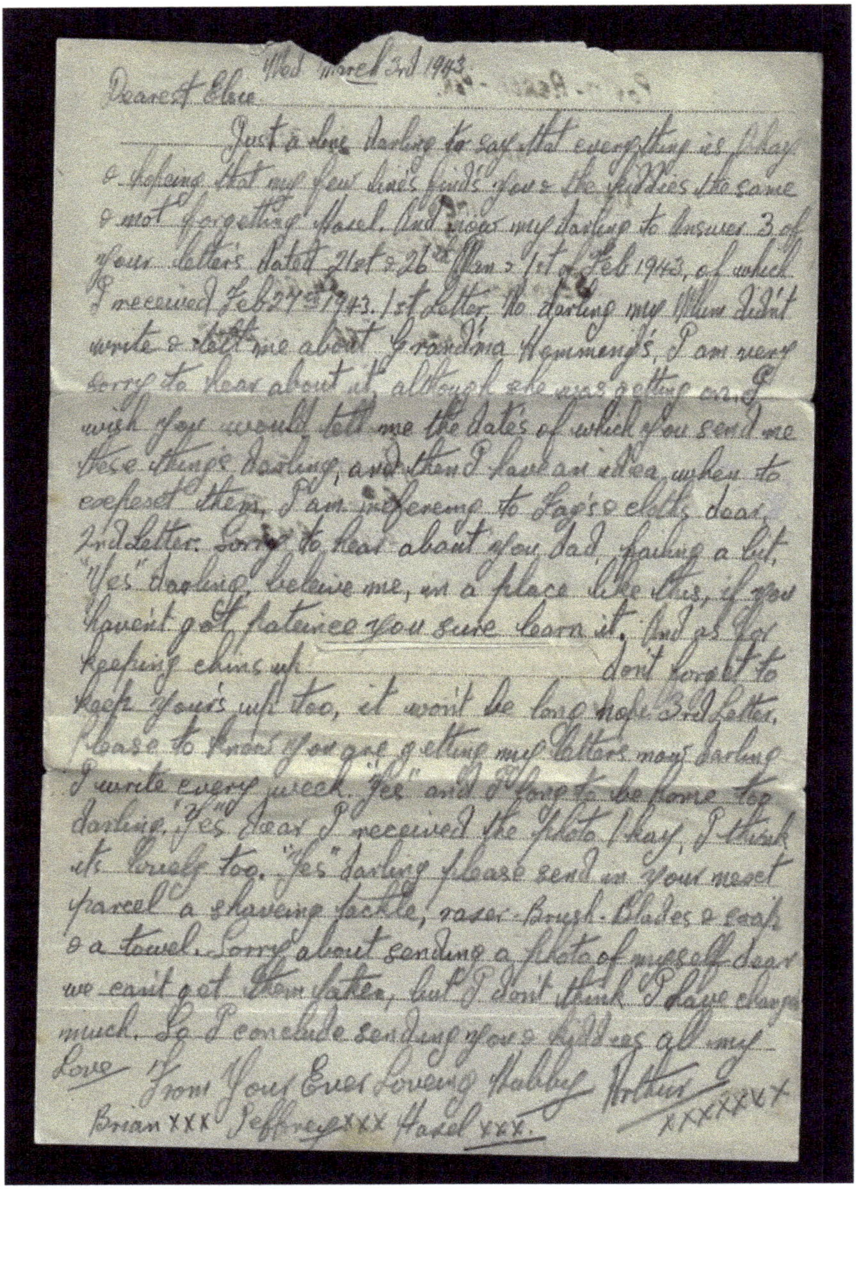
Wed March 3rd 1943.
Dearest Elsie
Just a line darling to say that everything is Okay
& hopeing that my few lines finds you & the kiddies the same
& not forgetting Hazel. And now my darling to answer 3 of
your letters dated 21st & 26th Jan & 1st of Feb 1943, of which
I received Feb 27th 1943. 1st Letter. No darling my Mam didn't
write & tell me about Grandma Hemmings, I am very
sorry to hear about it although she was getting on. I
wish you would tell me the dates of which you send me
these things darling, and then I have an idea when to
expesct them. I am refereing to Fay's clothes dear.
2nd Letter. Sorry to hear about you Dad, failing a bit,
"Yes" darling, beleive me, in a place like this, if you
haven't got paitence you sure learn it. And as for
keeping chins up don't forget to
keep your's up too, it won't be long hope. 3rd Letter.
Please to know you are getting my letters may darling
I write every week. "Yes" and I long to be home too
darling. "Yes" dear I received the Photo Okay, I think
its lovely too. "Yes" darling please send in your next
parcel a shaving tackle, raser-Brush-Blades & soap
& a towel. Sorry about sending a photo of myself dear
we can't get them taken, but I don't think I have changed
much. So I conclude sending you & kiddies all my
Love
From Your Ever Loving Hubby Arthur
XXXXXXX
Brian XXX Jeffrey XXX Hazel XXX.

Wed March 3rd, 1943

Dearest Elsie

Just a line darling to say that everything is okay and hoping that my few lines find you and the kiddies the same and not forgetting Hazel. And now my darling to answer 3 of your letters dated 21st, 26th and 1st Feb 1943 of which I received on February 27th, 1943. 1st Letter, no darling my mum didn't write and tell me about Grandad Hemmings, I am very sorry to hear, although he was getting old. I wish you would tell me the dates of which you send me these things darling and then I have an idea of when to expect these things. I am referring to fags and clothes dear. 2nd Letter, sorry to hear about your dad failing a bit. "Yes" darling believe me in a place like this if you haven't got patience, you sure learn it. As for keeping your chin up don't forget to keep yours up too, it won't be long now. Letter 3, pleased to know you are getting my letters now darling I write every week. "Yes" I long to be home too darling. "Yes" I received the photo okay I think its lovely too. "Yes" darling please send in your next parcel a shaving tackle, razor and brush, blades, soap and a towel. Sorry about sending a photo of myself dear we can't get them taken, but I don't think I have changed much. So, I conclude sending you and kiddies all my love from your ever-loving Hubby Arthur xxxxxxx

Brian xxx

Jeffrey xxx

Hazel xxx

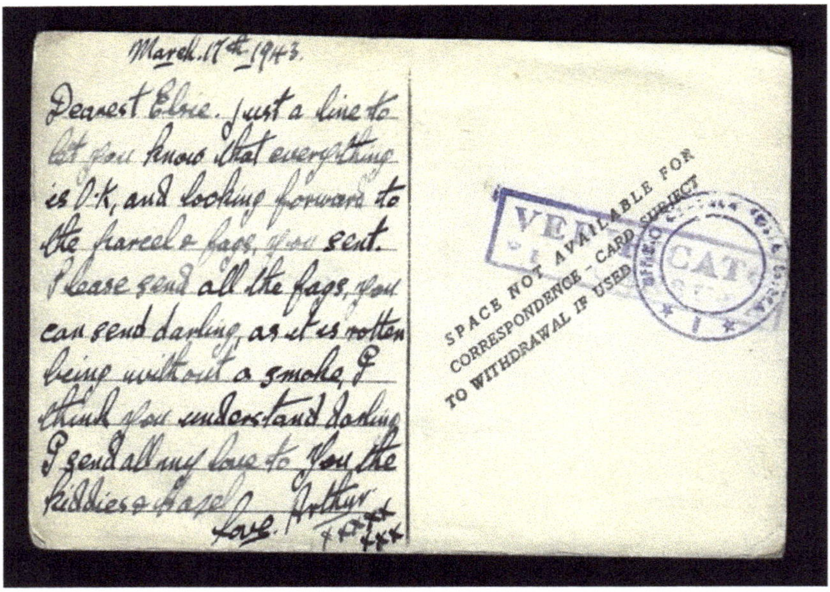

March 17th 1943

Dearest Elsie, just a line to let you know that everything is OK and looking forward to the parcel and fags you sent. Please send all he fags you can send darling as it is rotten being without a smoke. I think you understand darling. I send all my love to you the kiddies and Hazel. Love Arthur xxxxxxxx

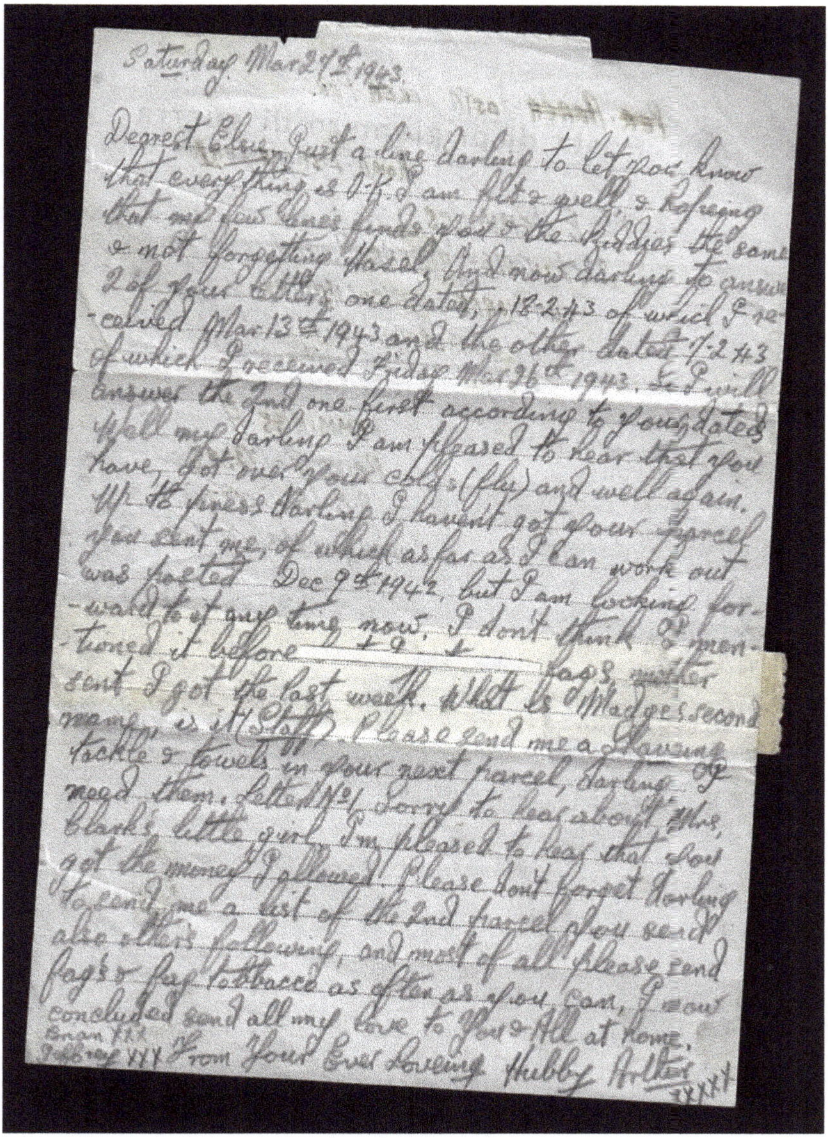

Saturday Mar 27th 1943.

Dearest Elsie, Just a line darling to let you know that everything is O.K. I am fit & well & hoping that my few lines finds you & the kiddies the same & not forgetting Hazel, And now darling to answer 2 of your letters one dated, 18-2-43 of which I re-ceived Mar 13th 1943 and the other dated 7-2-43 of which I received Friday Mar 26th 1943, So I will answer the 2nd one first according to you dates. Well my darling I am pleased to hear that you have got over your colds (flu) and well again. Up to press darling I haven't got your parcel you sent me, of which as far as I can work out was posted Dec 9th 1942, but I am looking for-ward to it any time now. I don't think I men-tioned it before [illegible] fags mother sent I got the last week. What is Madges second name, is it Staff. Please send me a Shaving tackle & towels in your next parcel, darling I need them. Letter No 1 Sorry to hear about Mrs Clarks little girl. I'm pleased to hear that you got the money I allowed. Please don't forget darling to send me a list of the 2nd parcel you send also others following, and most of all please send fags & fag tobacco as often as you can, I now concluded send all my love to you & All at home.

[illegible] XXX
[illegible] XXX From Your Ever Loving Hubby Arthur XXXX

Saturday Mar 27th 1943

Dearest Elsie, Just a line darling to let you know that everything is OK. I am fit and well and hoping my few lines finds you and the kiddies the same and not forgetting Hazel. And now darling to answer 2 of your letters one dated 18.2.43 of which I received Mar 13th, 1943, and the other dated 7.2.43 of which I received Friday morning, March 26th 1943. So, I will answer the 2nd one first according to your dates well my darling I am pleased to hear that you have got over your cold (flu) and well again. Up to press darling I haven't got your parcel you sent me, of which as far as I can work out was posted Dec 9th 1942, but I am looking forward to it any time now. I don't think I mentioned it before, but I got fags mother sent, I got them last week. What is Madges second name is it Staff. Please send me a shaving tackle and towels in your next parcel darling, I need them. Letter No 1 sorry to hear about Mrs Clarks little girl. I am pleased to hear that you got the money I allowed. Please don't forget darling to send me a list of the 2nd parcel you send also others following and most of all please send fags, fag tobacco as often as you can. I now concluded send all my love to you and all at home. From your ever loving Hubby Arthur xxxxx

Brian xxx

Jeffrey xxx

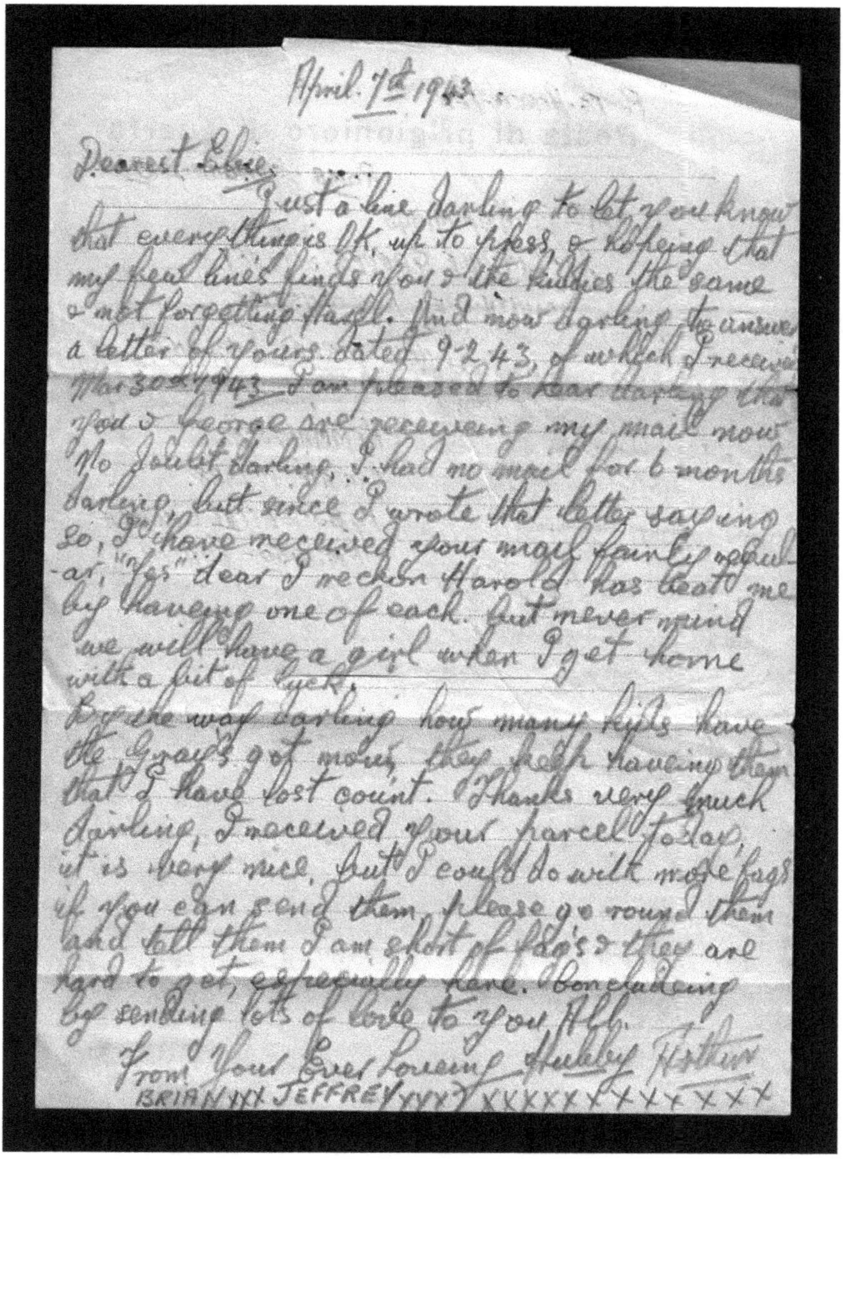

April 7th 1943

Dearest Elsie

Just a line darling to let you know
that everything is OK, up to press, & hoping that
my few lines finds you & the kiddies the same
& not forgetting Hazel. And now darling to answer
a letter of yours dated 9-2-43, of which I received
Mar 30th 1943. I am pleased to hear darling that
you & George are receiving my mail now
No doubt darling, I had no mail for 6 months
darling, but since I wrote that letter saying
so, I have received your mail fairly regul-
-ar, "Yes" dear I reckon Harold has beat me
by having one of each. But never mind
we will have a girl when I get home
with a bit of luck.
By the way darling how many kids have
the Gray's got now, they keep having them
that I have lost count. Thanks very much
darling, I received your parcel today,
it is very nice, but I could do with more fags
if you can send them, please go round them
and tell them I am short of fag's & they are
hard to get, especially here. Concluding
by sending lots of love to you All.

From Your Ever Loving Hubby Arthur

BRIAN XXX JEFFREY XXX XXXXXXXXXXXX

April 7th 1943

Dearest Elsie

Just a line darling to let you know that everything is OK up to press and hoping that my few lines find you and the kiddies the same and not forgetting Hazel. And now darling to answer a letter of yours dated 9.2.43, of which I received Mar 30th, 1943. I am pleased to hear darling that you and George are receiving my mail now no doubt darling. I had no mail for 6 months darling but since I wrote that letter saying so I have received your mail fairly recent. "Yes" dear I reckon Harold has got me by having one of each but never mind we will have a girl when I get home with a bit of luck.

By the way darling how many kids have the Graves got now, they keep having them that I have lost count. Thanks very much darling I received your parcel today it is very nice, but I could do with more fags if you can send them. Please go round them and tell them I am short of fags and they are hard to get especially here. I am concluding by sending lots of love to you and you all, from your ever loving Hubby Arthur xxxxxxxxxxxx

Brian xxx

Jeffrey xxx

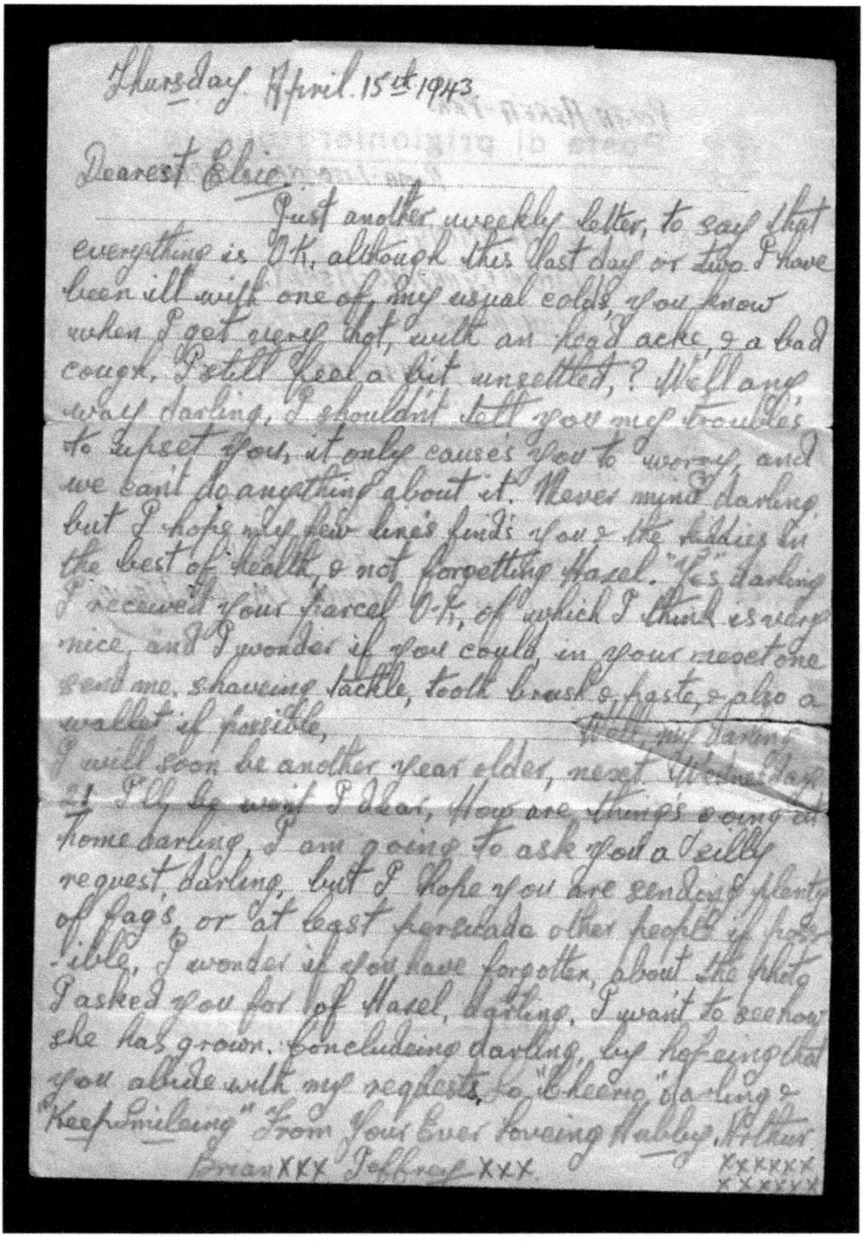

Thursday. April. 15th 1943.

Dearest Elsie

Just another weekly letter, to say that everything is O.K. although this last day or two I have been ill with one of my usual colds, you know when I get very hot, with an head ache, & a bad cough. I still feel a bit unsettled, ? Well any way darling, I shouldn't tell you my trouble's to upset you, it only cause's you to worry, and we can't do anything about it. Never mind darling, but I hope my few line's finds you & the kiddies in the best of health, & not forgetting Hasel. "Yes" darling I received your parcel O.K., of which I think is very nice, and I wonder if you could, in your next one send me, shaveing tackle, tooth brush & paste, & also a wallet if possible, Well my darling I will soon be another year older, next Wednesday 21 I'll be [illegible] dear, How are thing's going at home darling, I am going to ask you a silly request, darling, but I hope you are sending plenty of fag's, or at least persuade other people if possible, I wonder if you have forgotten about the photo I asked you for of Hasel, darling. I want to see how she has grown. Concluding darling, by hopeing that you abide with my requests, so "Cheerio" darling & "Keep Smileing" From Your Ever loveing Hubby, Arthur

Brian XXX Jeffrey XXX XXXXXX XXXXXX

Thursday April 15th 1943

Dearest Elsie

Just another weekly letter to say that everything is OK, although this last day or two I have been ill with one of my usual colds. You know when I get very hot with a headache and a bad cough. I still feel a bit unsettled. Well anyway darling I shouldn't tell you my troubles to upset you it only causes you to worry and we can't do anything about it. Never mind darling but I hope my few lines finds you and the kiddies in the best of health and not forgetting Hazel. "Yes" darling I received your parcel OK, of which I think is very nice and I wonder if you could in your next one send me shaving tackle, toothbrush and paste and also a wallet if possible. Well, my darling I will soon be another year older next Wednesday 21, I will be wont I dear. How are things going at home darling, I hope you are sending plenty of fags or at least persuade other people if possible. I wonder if you have forgotten about the photo I asked for of Hazel darling. I want to see how she has grown. Concluding darling by hoping that you abide with my request. So cheerio darling and keep smiling. From your ever loving Hubby Arthur xxxxxxxxxxxx

Brian xxx

Jeffrey xxx

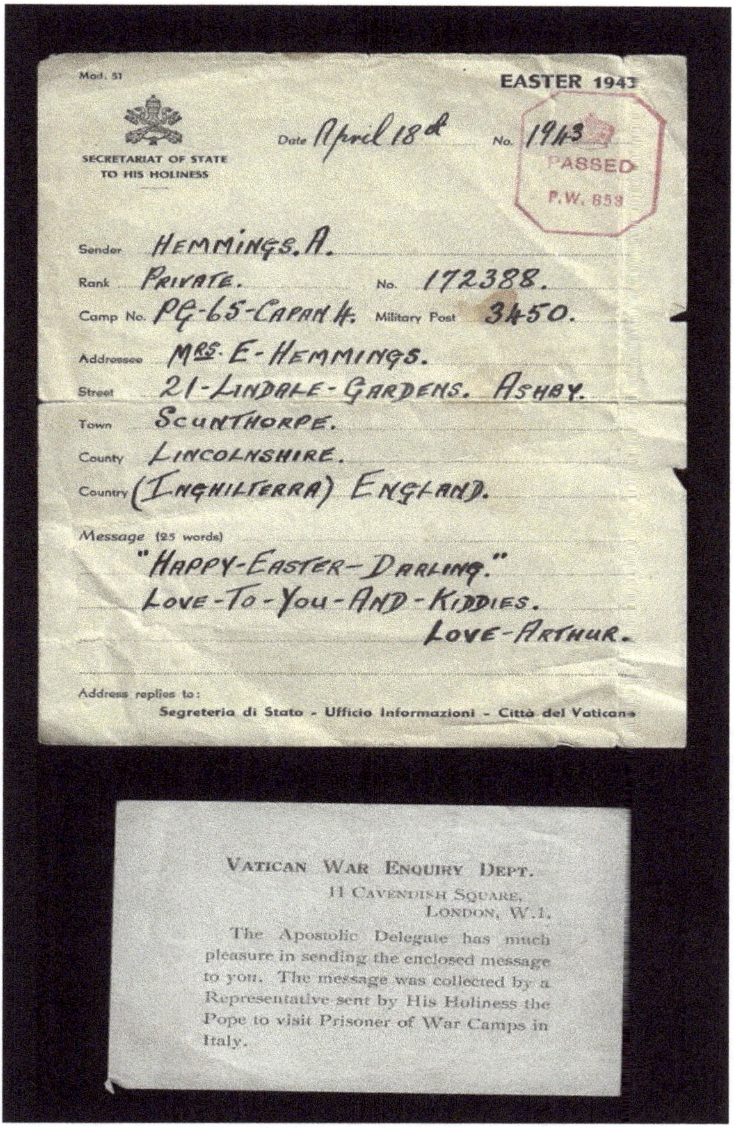

Mod. 51

EASTER 1943

SECRETARIAT OF STATE TO HIS HOLINESS

Date April 18th No. 1943

PASSED P.W. 853

Sender HEMMINGS. A.

Rank PRIVATE. No. 172388.

Camp No. PG-65-CAPAN IV. Military Post 3450.

Addressee MRS E-HEMMINGS.

Street 21-LINDALE-GARDENS. ASHBY.

Town SCUNTHORPE.

County LINCOLNSHIRE.

Country (INGHILTERRA) ENGLAND.

Message (25 words)

"HAPPY-EASTER-DARLING."
LOVE-TO-YOU-AND-KIDDIES.
LOVE-ARTHUR.

Address replies to:
Segreteria di Stato - Ufficio Informazioni - Città del Vaticano

VATICAN WAR ENQUIRY DEPT.
11 CAVENDISH SQUARE,
LONDON, W.1.

The Apostolic Delegate has much pleasure in sending the enclosed message to you. The message was collected by a Representative sent by His Holiness the Pope to visit Prisoner of War Camps in Italy.

This is a really special letter. The Priests, from the Vatican used to visit the prisoners on special occasions and write these letters. They would be taken to the Vatican and posted to the UK

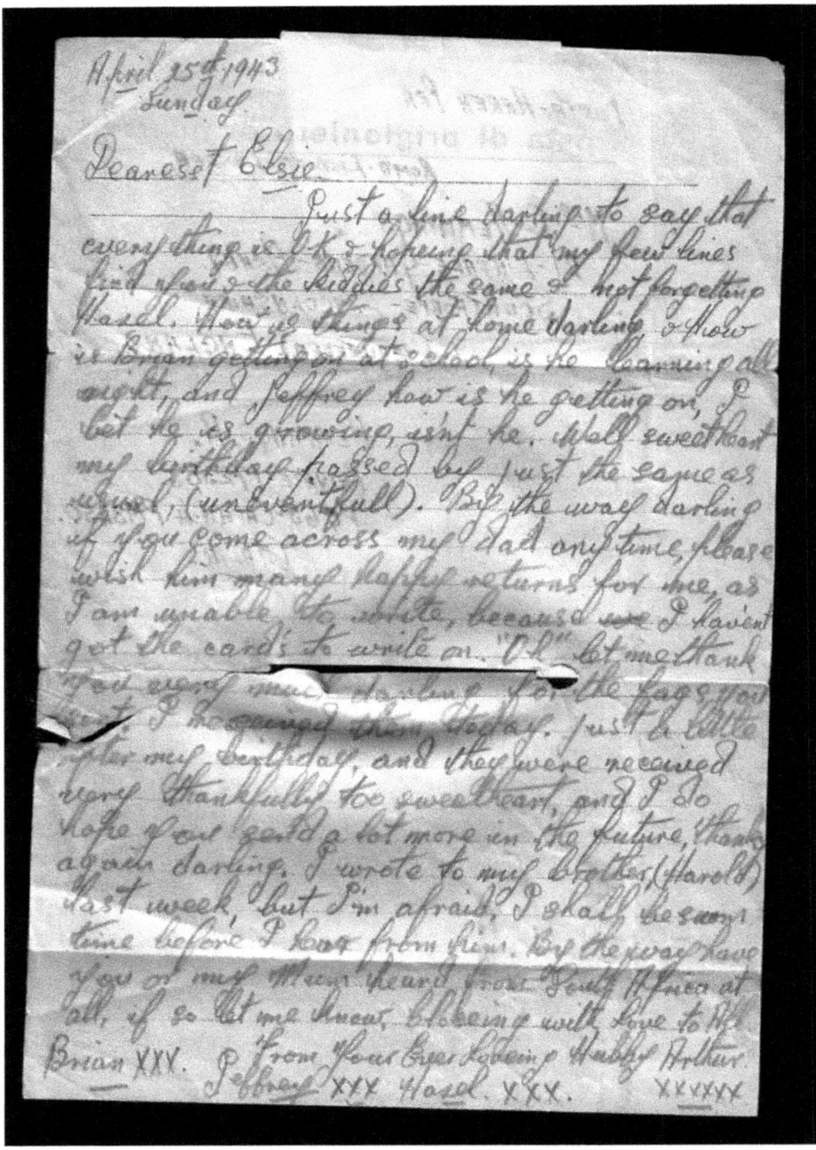

April 25th 1943
Sunday

Dearest Elsie

Just a line darling to say that everything is O.K. & hoping that my few lines find you & the kiddies the same & not forgetting Hazel. How is things at home darling & how is Brian getting on at school, is he learning all right, and Jeffrey how is he getting on, I bet he is growing, isn't he. Well sweetheart my birthday passed by just the same as usual, (uneventfull). By the way darling if you come across my dad any time, please wish him many happy returns for me, as I am unable to write, because I haven't got the cards to write on. "Oh" let me thank you very much darling for the fags you sent, I received them today, just a little after my birthday, and they were received very thankfully too sweetheart, and I do hope you send a lot more in the future, thanks again darling. I wrote to my brother (Harold) last week, but I'm afraid I shall be some time before I hear from him. By the way have you or my Mum heard from South Africa at all, if so let me know, closing with love to All

Brian XXX. From Your Ever Loving Hubby Arthur.
Jeffrey XXX Hazel. XXX. XXXXXX

April 25th 1943 Sunday

Dearest Elsie

Just a line darling to say that everything is Ok and hoping that my few lines find you and the kiddies the same and not forgetting Hazel. How are things at home darling and how is Brian getting on at school, is he learning alright and Jeffrey how is he getting on, I bet he is growing isn't he. Well sweetheart my birthday passed by just the same as usual (uneventful). By the way darling if you come across my dad anytime please wish him many happy returns for me as I am unable to write because I haven't got the cards to write on. "Ok" let me thank you very much darling for the fags you sent, I received them today. Just a little after my birthday and they were received very thankfully too sweetheart and I do hope you send a lot more in the future. Thanks again darling. I wrote to my brother (Harold) last week but I'm afraid I think it will be some time before I hear from him. By the way have you or my mum heard from South Africa at all, if so let me know. Closing with love to all from your ever loving Hubby Arthur xxxxxx

Brian xxx

Jeffrey xxx

Hazel xxx

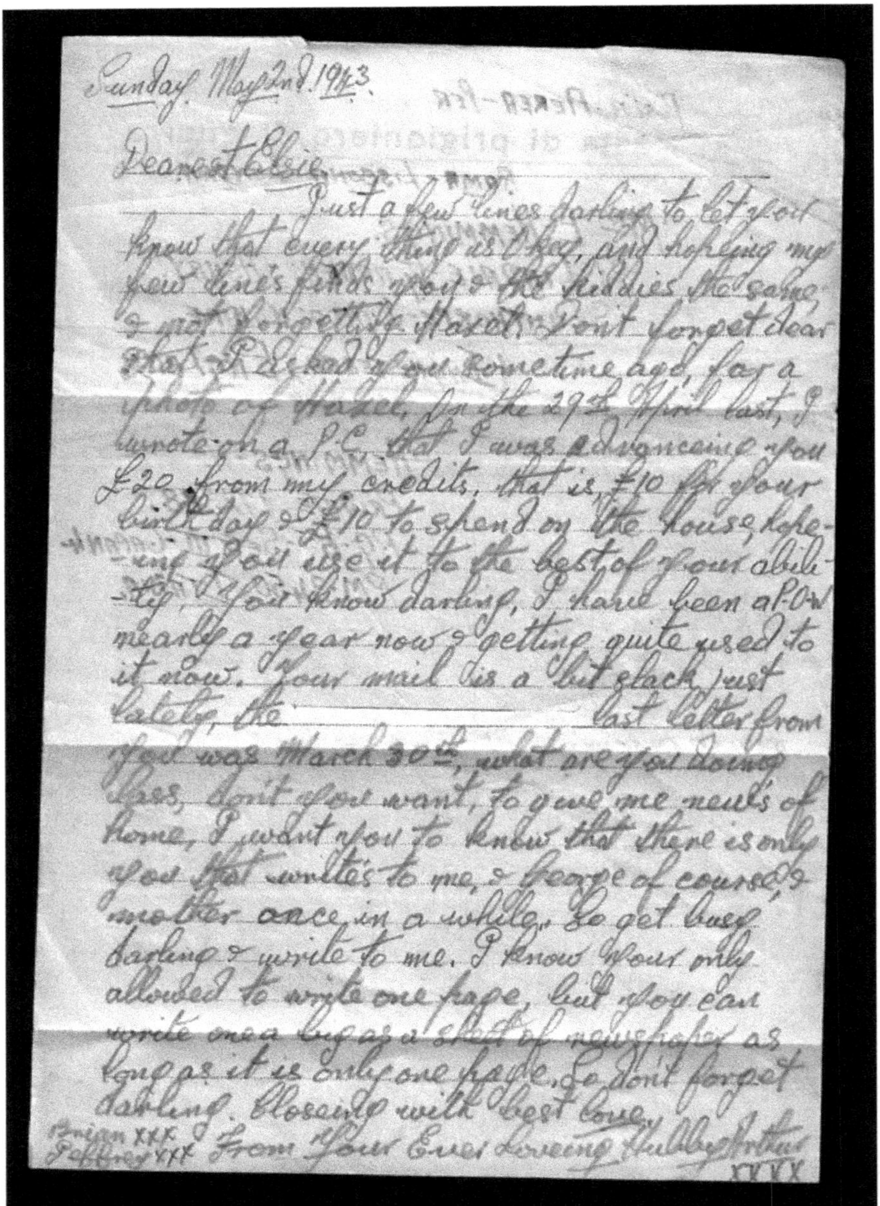

Sunday. May 2nd 1943.

Dearest Elsie

Just a few lines darling to let you know that every thing is O.Key, and hoping my few line's finds you & the kiddies the same, & not forgetting Hazel. Don't forget dear that I asked you sometime ago, for a photo of Hazel. On the 29th April last, I wrote on a P.C. that I was advancing you £20 from my credits, that is, £10 for your birthday & £10 to spend on the house, hopeing you use it to the best of your ability. You know darling, I have been at O.W. nearly a year now & getting quite used to it now. Your mail is a bit slack just lately, the last letter from you was March 30th, what are you doing Dear, don't you want, to give me news of home, I want you to know that there is only you that writes to me, & George of course, & mother once in a while, so get busy darling & write to me. I know your only allowed to write one page, but you can write me a big as a sheet of newspaper as long as it is only one page, so don't forget darling. Closeing with best love,

Brian XXX
Jeffrey XXX

From Your Ever Loveing Hubby Arthur
XXXX

Sunday May 2nd, 1943

Dearest Elsie,

Just a few lines darling to let you know that everything is okay and hoping my few lines find you and the kiddies the same and not forgetting Hazel. Don't forget dear that I asked you some time ago for a photo of Hazel. On the 29th of April last time I wrote on a PC, that I was sending you £20 from my credits that is £10 for your birthday and £10 to spend on the house, hoping you use it to the best of your ability. You know darling I have been a POW nearly a year now getting quite used to it now. Your mail is a but slack just lately the last letter from you was March 30th what are you doing lass, don't you want to give me news of home. I want you to know that there is only you that writes to me and George of course and mother once in a while. Do get busy darling and write to me. I know your only allowed to write one page but you can write one big sheet of newspaper as long as it is only one page. So don't forget darling. Closing with best love from your ever-loving Hubby Arthur xxxx

Brian xxx

Jeffrey xxx

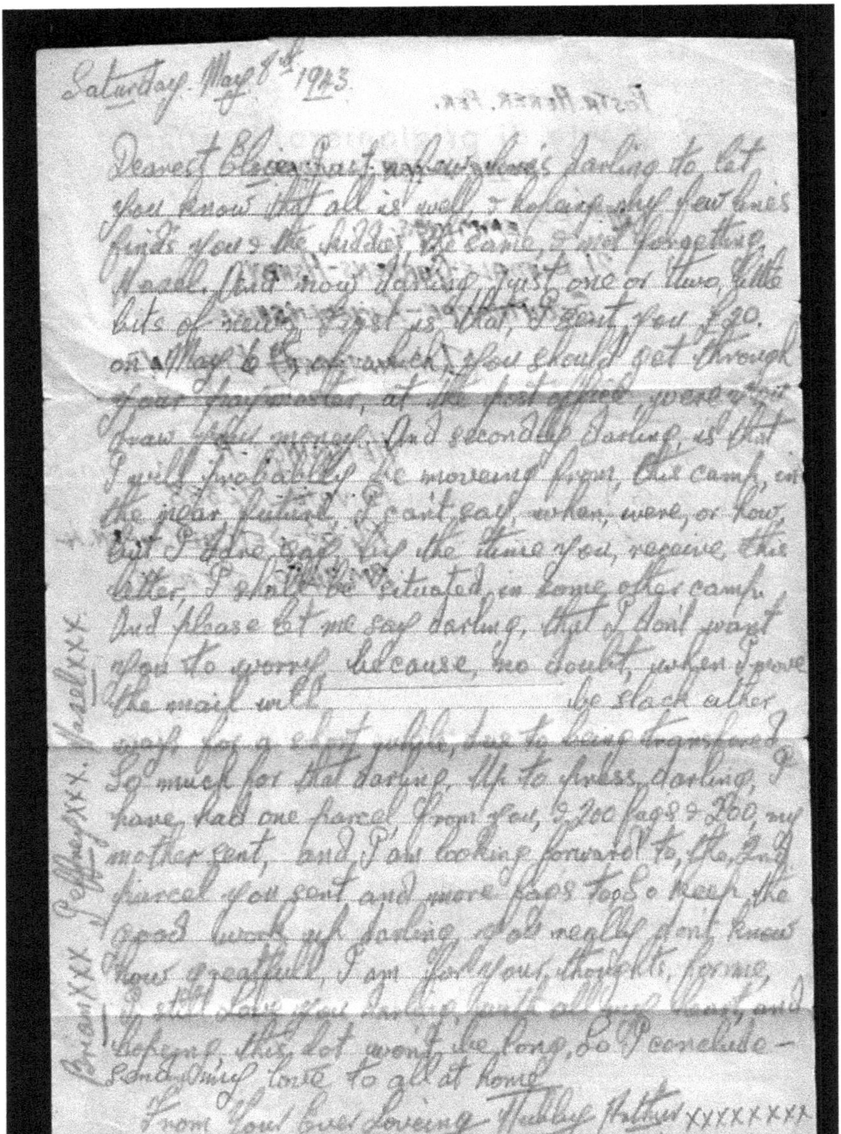
Saturday May 8th 1943.

Dearest Eleanor Just a few lines darling to let you know that all is well, & hoping this few lines finds you & the kiddies the same, & not forgetting Hazel. And now darling just one or two little bits of news first is that, I sent you £20. on May 6 [illegible] which you should get through your paymaster, at the post office, were you draw your money. And secondly darling is that I will probably be moving from, this camp, in the near future, I can't say, when, were, or how, but I dare say by the time you, receive this letter, I shall be situated in some other camp. And please let me say darling, that I don't want you to worry because, no doubt, when I move the mail will be slack either way for a short while, due to being transfered. So much for that darling, Up to press, darling, I have had one parcel from you, & 200 fags & 200, my mother sent, and I am looking forward to, the 2nd parcel you sent and more fags too. So keep the good work up darling, as I really don't know how greatfull I am for your thoughts, for me, I still love you darling with all my heart, and hoping this lot won't be long, so I conclude — sending my love to all at home

From Your Ever Loveing Hubby [illegible] XXXXXXXX

Brian XXX Geoffrey XXX. Hazel XXX.

Saturday May 8th 1943

Dearest Elsie just a few lines darling to let you know that all is well and hoping my few lines finds you and the kiddies the same. Now darling just one or two little bits of news first is that I sent you £20 on May 6th of which you should get through your paymaster at the post office where you draw your money. And secondly darling is that I will probably be moving from this camp in the near future, I can't say when, where or how but I dare say by the time you receive this letter I shall be situated in some other camp. And please let me say darling that I don't want you to worry because no doubt when I move the mail will be slack anyway for a short while, due to being transferred. So much for that darling. Up to press darling have had one parcel from you and 200 fags and 200 my mother sent and I am looking forward to the 2nd parcel you sent and move fags too. Keep the good work up darling you really don't know how grateful I am for your thoughts for me. I still love you darling with all my heart and hoping this lot won't be long. So, I conclude sending my love to all at home. From your every loving Hubby Arthur xxxxxxxx

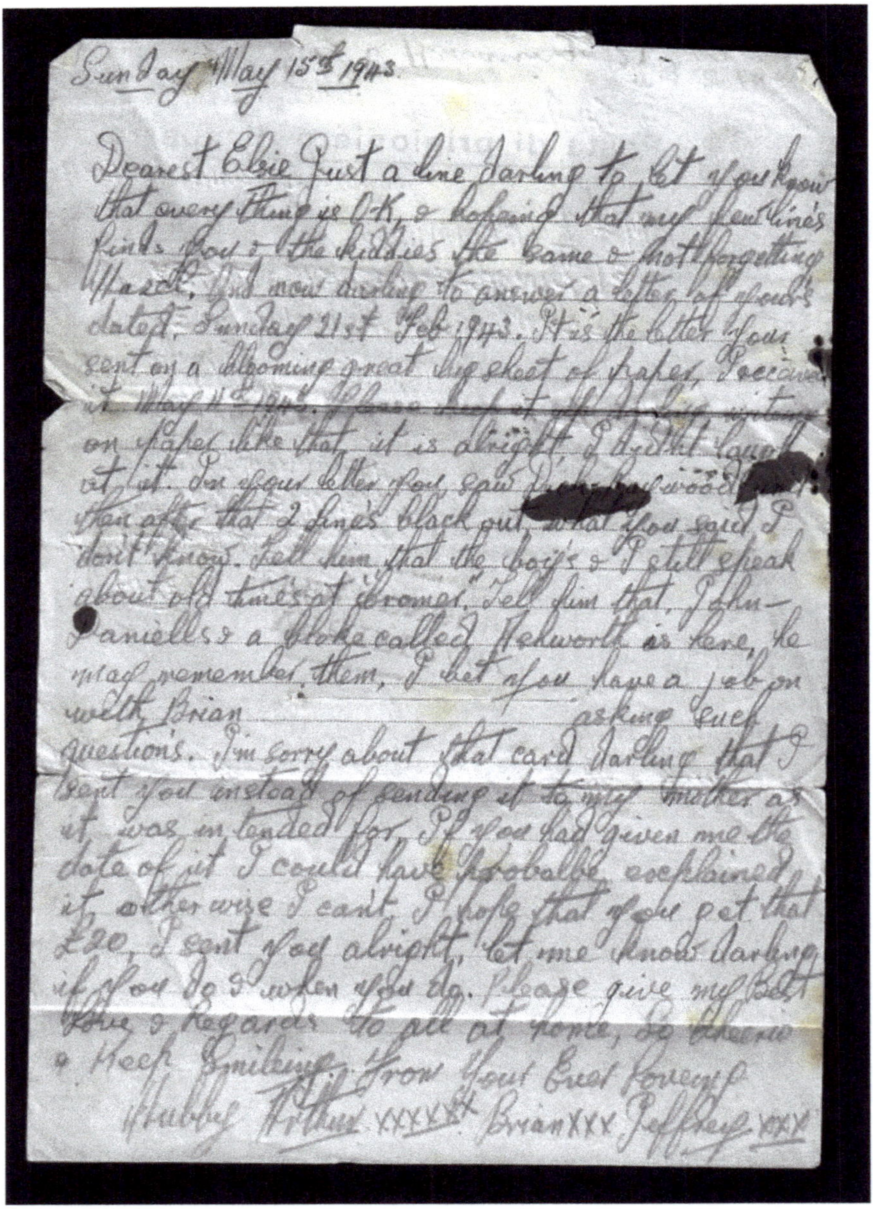

Sunday May 15th 1943.

Dearest Elsie Just a line darling to let you know that every thing is O.K. & hoping that my few line's finds you & the kiddies the same & not forgetting [illegible]. And now darling to answer a letter of your's dated Sunday 21st Feb 1943. It's the letter you sent on a blooming great big sheet of paper, I received it May 11th 1943. [illegible] on paper like that it is alright I didn't laugh at it. In your letter you said [illegible]wood [illegible] then after that 2 line's black out, what you said I don't know. Tell him that the boy's & I still speak about old times at Cromer. Tell him that John Daniells & a bloke called [illegible]worth is here, he may remember them, I bet you have a job on with Brian asking such question's. I'm sorry about that card darling that I sent you instead of sending it to my mother as it was in tended for. If you had given me the date of it I could have probably explained it, other wise I can't. I hope that you got that £20, I sent you alright, let me know darling if you do & when you do. Please give my Best love & regards to all at home, So cheerio & Keep Smiling from your Ever loving

Hubby Arthur XXXXX Brian XXX Jeffrey XXX

Sunday May 15th 1943

Dearest Elsie, just a line darling to let you know that everything is OK and hoping that my few lines finds you and the kiddies the same and not forgetting Hazel. And now darling to answer a letter of yours dated Sunday 21st Feb 1943. It's the letter you sent on a blooming great big sheet of paper, received it May 11th 1943. Please keep it up darling writing on paper like that it is alright I didn't laugh at it. In your letter you saw Dick Haywood and then after that 2 lines black out what you said I don't know. Tell him that the boys and I still speak about old times at Cromer. Tell him that John Daniells and a bloke called Ashworth is here, he may remember them. I bet you have a job on with Brian asking such questions. I am sorry about the card darling that I sent you instead of it to my mother as it was intended for. If you had given me the date of it I could have probably explained it otherwise I cant. I hope that you got that £20 I sent you alright, let me know darling when you do. Please give my love and regards to all at home. So cheerio and keep smiling from your ever loving Hubby Arthur xxxxxx

Brian xxx

Jeffrey xxx

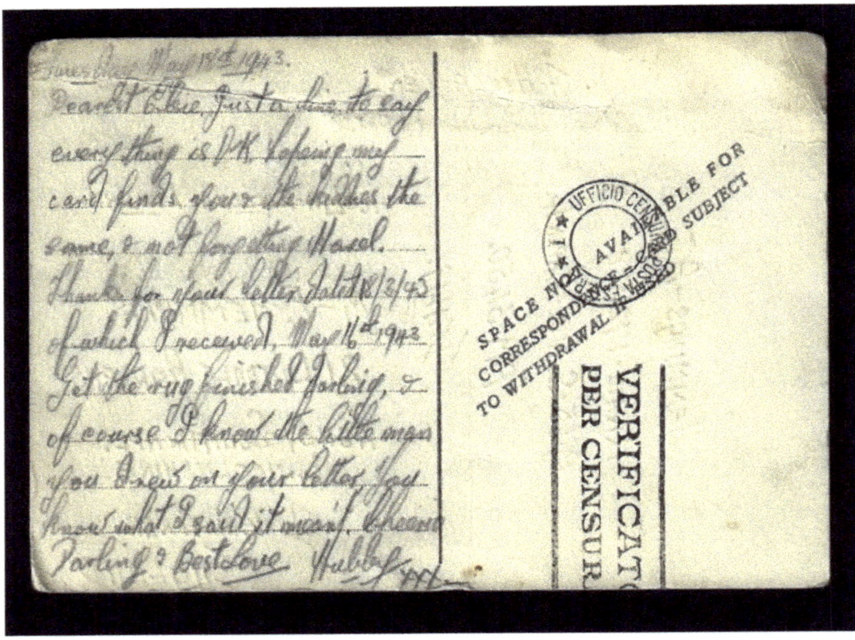

Thursday May 18th, 1943

Dearest Elsie, just a line to say everything is OK hoping my card finds you and the kiddies the same and not forgetting Hazel. Thanks for your letter dated 18/3/43 of which I received May 16th, 1943. Get the rug finished darling and of course I know the little man you drew on your letter you know what I said it meant. Cheerio darling and best love Hubby xxx

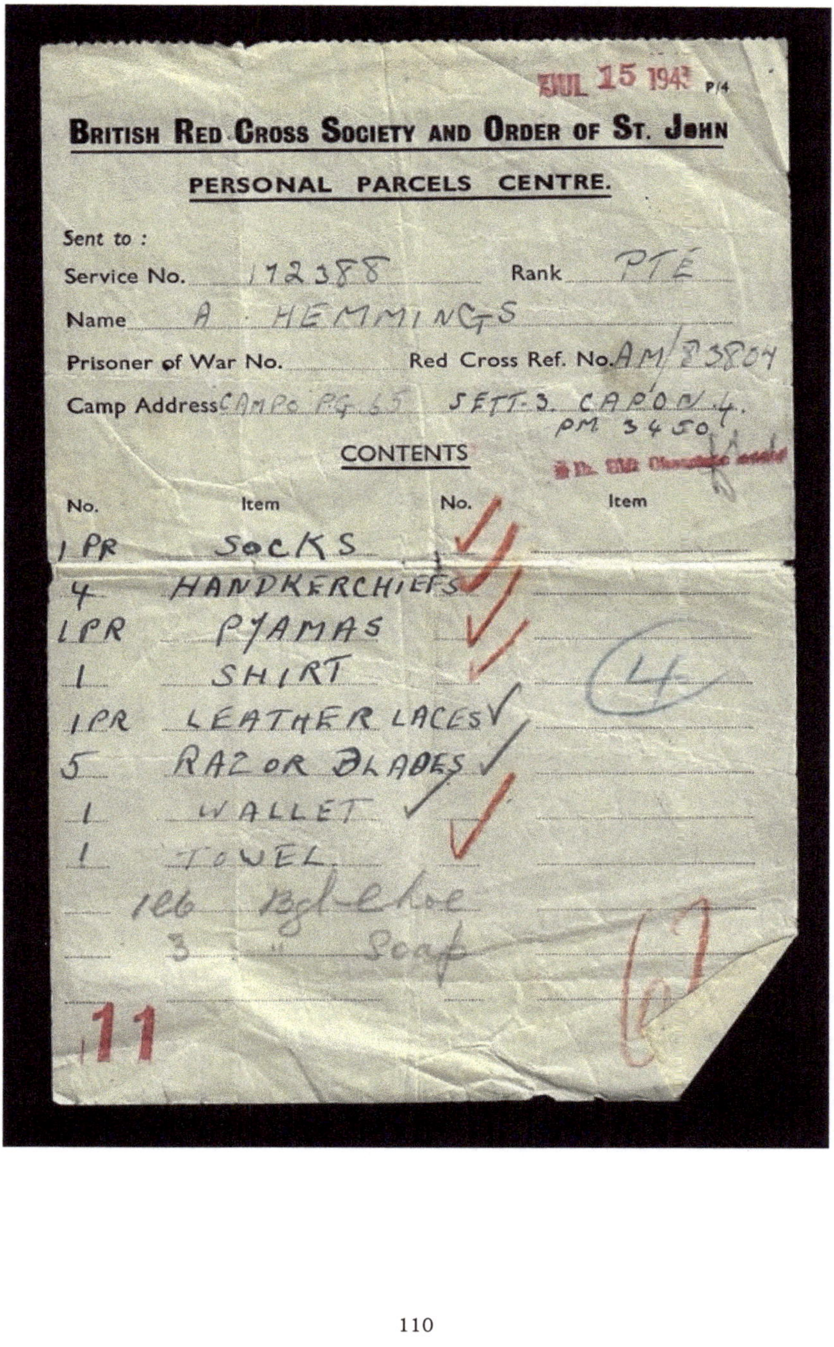

JUL 15 1943 P/4

BRITISH RED CROSS SOCIETY AND ORDER OF ST. JOHN

PERSONAL PARCELS CENTRE.

Sent to :

Service No. 172388 Rank PTE

Name A HEMMINGS

Prisoner of War No. Red Cross Ref. No. AM/83804

Camp Address CAMPO PG 65 SETT-3. CAPON 4. PM 3450

CONTENTS

No.	Item	No.	Item
1 PR	SOCKS		
4	HANDKERCHIEFS		
1 PR	PYAMAS		
1	SHIRT		
1 PR	LEATHER LACES		
5	RAZOR BLADES		
1	WALLET		
1	TOWEL		
1lb	Bgl. choc		
3	" Soap		

11

CHAPTER SEVEN

CAMPO 145

Set high in the mountains of central Italy lies a village so small and unassuming that most maps barely mark it.

Montorio al Vomano, in the district of Campotosto, province of L'Aquila, seems at first glance like any other quiet Italian hamlet.

It clings to the slopes above Lake Campotosto, a scattering of houses of stone and plaster, a church tower watching over winding alleys, all of it hemmed in by ridges of jagged mountains. In winter, snow buries the roofs.

In summer, the slopes glow gold beneath a wide, open sky. It is beautiful, isolated, and serene.

Today, the place keeps its secrets well.

Ask the locals, and most will shrug if you mention what once stood here.

They will talk about the lake, the mountains, the hard lives carved from thin soil, and the way young people drifted away to the cities.

But about the camp, they shake their heads.

Nothing remains. Not a plaque. Not a ruin.

Not even a story handed down in whispers.

It is as if Campo 145 never existed.

There is a war memorial in the centre of the village, but no mention, that I could see, of the POW camp.

When I explained why I was there, their expressions changed.

I told them about my grandad, about Tobruk, about the nine days in the desert, about Italy, about Campo 65, and then Campo 145.

I told them he had been a prisoner here, that he had walked this land not as a tourist but as a captive.

Some listened quietly. Some were surprised. A few asked questions.

For many, it was the first time they had heard that Allied soldiers had once been held in their mountains.

In that moment, I realised something important.

If there is no marker, no stone, no official memory, then the stories themselves become the memorial.

By standing there and speaking his name, I was giving the place back its history. I was reminding the land, and perhaps the people, that it had once witnessed fear, endurance, and courage.

Campo 145 may have disappeared from the landscape, but it has not disappeared from our family.

And as long as his story is told, it never truly will.

Campo 145 was used from July 1942 to September 1943.

The camp was small by Italian standards, holding no more than three hundred men.

A mere fragment of the tens of thousands imprisoned across the country.

The prisoners lived in wooden huts, a stark contrast to the stone barracks of larger camps like Campo 65. These huts were temporary structures, erected quickly and cheaply, offering little protection from the mountain climate.

In winter, snow piled high against the walls, and icy winds found their way through every crack in the timber.

Boots froze stiff overnight. Blankets grew damp and never quite dried.

In summer, the huts trapped heat, turning stifling by day and only slightly more bearable at night.

Men slept packed close together, sharing breath, sharing silence.

Campo 145 was a working camp.

Each morning, prisoners were marched out under guard into the surrounding countryside. Some were sent into the forests to cut timber.

Others dug ditches, hauled supplies, or worked on nearby farms.

The labour was exhausting and repetitive, yet it carried one small advantage. It brought contact with the outside world.

A farmer's nod. A whispered warning. An apple pressed into a hand when no one was looking. For men starved of information as much as food, these moments mattered.

The camp held a mix of nationalities: Serbian soldiers captured during the Balkan campaigns, South Africans taken in North Africa, and New Zealanders who had fought far away in desert battles.

When Campo 65 closed in July 1943, prisoners were scattered across smaller camps, uprooted once again, and moved with little explanation.

My grandad was one of them, sent north into the mountains, away from stone huts, into a camp that felt quieter but no less uncertain.

Why he was sent to Montorio al Vomano remains unclear.

Perhaps numbers needed balancing.

Perhaps he was considered fit for labour.

Perhaps it was simply a chance.

But the move marked a turning point.

Camp 145 was isolated, far from major transport routes, perched in terrain that was difficult to control. Without intending it, the Italian authorities had placed him somewhere that escape would one day be possible.

That day came sooner than anyone expected.

On the evening of 8 September 1943, news filtered into the camp that Italy had surrendered.

Marshal Pietro Badoglio's announcement declared that Italy had signed an armistice with the Allies, effectively ending its participation in the war.

The purpose of the armistice was to halt fighting on Italian soil, prevent further devastation, and allow Allied forces to advance more quickly.

For prisoners of war, it created a chaotic mixture of opportunity and danger.

Authority collapsed almost overnight.

Mussolini was gone.

Italy had changed sides.

German forces were moving quickly to take control, but in many places, including Montorio al Vomano, they had not yet arrived.

Italian guards did not know what to do.

Some vanished overnight, slipping back to their villages, discarding uniforms, no longer willing to stand watch over prisoners when the war itself seemed to have shifted beneath their feet.

Others stayed but stopped enforcing rules. The gates felt less guarded.

Inside the huts, I imagined rumours spreading in low voices:

"Italy's finished."

"The Germans will take us north."

"They'll send us to Germany if we stay."

"This is our chance."

Across Italy, Senior British Officers urged men to remain where they were, following earlier Allied instructions to "stay put" until help arrived.

Many obeyed. Thousands were later recaptured when German troops swept in to take control.

But in remote camps like Campo 145, the breakdown came first. Authority dissolved before the replacement arrived.

For a brief, dangerous window, the camp existed in limbo, no longer a prison, not yet reclaimed.

I am not certain, but I am sure my grandad did not hesitate.

Quiet by nature, he said little to anyone, but when the moment came, I imagine he made his decision.

In early October 1943, with German patrols still tightening their grip elsewhere, he slipped away from Campo 145 and into the mountains.

I do not know if he was alone when he escaped or with other prisoners, but according to his letters, he took the chance to escape.

There was no dramatic exit. No final glance back. Just a careful step beyond what had once been forbidden. A walk uphill into trees, rock, and open sky.

By September and October, when my grandad escaped, the mountains around *Amici* had lost the last traces of summer softness.

They were not especially high, but they remained unforgiving. Tree-covered ridges rose sharply from the valleys, their slopes scattered with loose rubble and jagged rocks that made every step uncertain.

Paths were scarce, often no more than narrow goat tracks winding along steep drops and through thick forests of pine, beech, and chestnut.

In places, the trees thinned to reveal open pastures, now rougher and fading, dotted with scrub and wild herbs turning with the season.

Water was harder to come by.

The streams that ran through the ravines were fewer now, no longer fed by melting snow, and never something a man could rely on.

What water there was ran cold enough to numb the body within minutes.

The wind carried a sharper edge, cutting through the valleys with the scent of pine and damp earth.

Villages remained scarce, tucked away in the folds of the mountains. Some inhabitants still risked everything to help those on the run; others, wary of German patrols, kept their doors firmly closed.

In this terrain, navigation was everything. Every ridge, every turn in the valley, could mean the difference between escape and running into the Germans.

It is in this terrain that my grandad moved, step by cautious step.

The mountains were his ally and his adversary, offering concealment, fresh water in hidden streams, shelter beneath trees or rocks, but never allowing a moment of ease.

He would have had to read the land as instinctively as he breathed, balancing hunger, exhaustion, and fear against the need to keep moving, keep climbing, keep surviving.

I have been unable to determine what happened in the 3 weeks he escaped, but this is how I imagine my grandad's escape would have been like:-

Three Weeks on the Run

Week One: Into the Wild

The first night, he walked without stopping. Boots scraping stone, breath steaming in the cold, nerves hotter than hunger. Behind him, the camp shrank into darkness.

Ahead, ridges loomed like walls. Forest swallowed sound. Every snap of a twig could mean discovery.

He was not entirely alone at first. One of the Serbian prisoners who had fled alongside him whispered, hoarsely:

"We head south. Always south. Capito?"

Grandad only nodded. They climbed goat tracks, stones tumbling underfoot. But dawn came, and with it, the reality that groups were easier to spot. They parted silently, each disappearing into the forest.

From then on, he was alone.

Sleep was shallow, stolen in hollows beneath rocks. Hunger came quickly; for two days, he ate nothing.

On the third, he found wild berries clinging to a bush, their sweet and sour juice staining his lips purple.

Water was rarer. Streams were thin, icy, sometimes dry.

Yet he moved, guided by instinct and the stars overhead. Polaris became his anchor.

Week Two: The Tightening Net

By the second week, German patrols scoured the valleys.

The villages were tense; helping an escaped prisoner could mean imprisonment or death. Yet some villagers risked everything.

One night, shivering outside a farmhouse, he dared to knock. The door creaked open, a lantern lighting the farmer's lined face.

"*Chi sei?*"

"*Prigioniero… inglese,*" my grandad managed.

The farmer hesitated, then stepped aside.

Inside, his wife ladled thin soup into a bowl. Steam rose like a blessing.

Grandad ate silently, eyes burning with exhaustion and gratitude.

The farmer leaned close and whispered:

"*You go south. Sempre a sud. Alleati …*"

He tapped his chest. "*Amici.Friend*"

At dawn, they pressed bread into his hand and sent him on his way, not wanting him to stay any longer than needed for fear of being caught.

Another night brought danger: voices on the road, harsh German commands cutting through the dark.

He flattened himself into the brush, hardly daring to breathe.

Boots crunched gravel; a torch beam swung past, lighting rocks inches from his face. "*Nichts. Weiter.*" "Nothing. Move on," so the patrol moved on.

Only then did he release the breath he had been holding.

Week Three: Toward Freedom

By the third week, he was a ghost. Beard ragged, eyes hollow, feet blistered inside worn boots.

Hunger dulled into numbness, but thirst never left.

Once, he collapsed beside a small mountain stream, scooping icy water into his mouth until his stomach cramped.

Lifting his head, he saw a shepherd watching silently from a ridge. They stared at each other for a long moment, then the shepherd raised a hand in a quiet salute and walked away.

One evening, cresting a ridge, he heard it before he saw it: the distant thud of artillery, the rising plume of smoke from battle.

Heart pounding, he stumbled down a valley track, spotting figures ahead. Fear shot through him. Were they Germans? Should he flee?

Then a voice called out, unmistakably in English:

"Hold it, don't shoot! He's one of ours!"

Hands steadied him, a canteen pressed into his hand, water spilling down his chin. Someone clapped his shoulder.

"You made it, mate. You're safe now."

Safe. After months behind wire, after weeks in the wilderness, he had survived and was free.

For me, standing in Montorio al Vomano, the air heavy with mountain stillness, it is not forgotten.

I imagine him there: stepping off the truck in June 1943, living through that summer in the huts, watching Italy change with the armistice, and finally slipping away into the mountains.

His footprints may have long since washed from the soil, but his story remains.

Men from a small camp in one hidden village. Yet from here they walked their way back to freedom, and history is written.

As soon as he was in safe hands, the Army sent a telegram to my nana, and my grandad wrote two letters to my her saying he was safe:

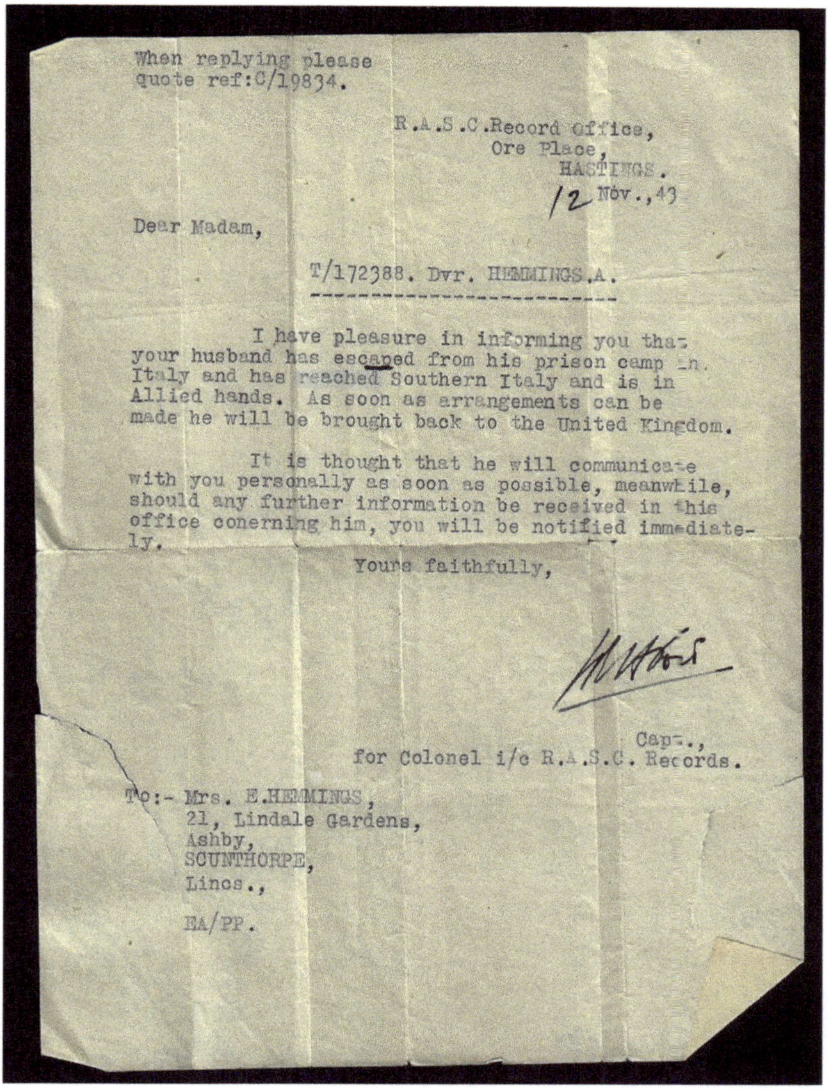

When replying please
quote ref:C/19834.

R.A.S.C.Record Office,
Ore Place,
HASTINGS.
12 Nov.,43

Dear Madam,

T/172388. Dvr. HEMMINGS.A.

I have pleasure in informing you that your husband has escaped from his prison camp in Italy and has reached Southern Italy and is in Allied hands. As soon as arrangements can be made he will be brought back to the United Kingdom.

It is thought that he will communicate with you personally as soon as possible, meanwhile, should any further information be received in this office conerning him, you will be notified immediately.

Yours faithfully,

Capt.,
for Colonel i/c R.A.S.C. Records.

To:- Mrs. E.HEMMINGS,
21, Lindale Gardens,
Ashby,
SCUNTHORPE,
Lincs.,

EA/PP.

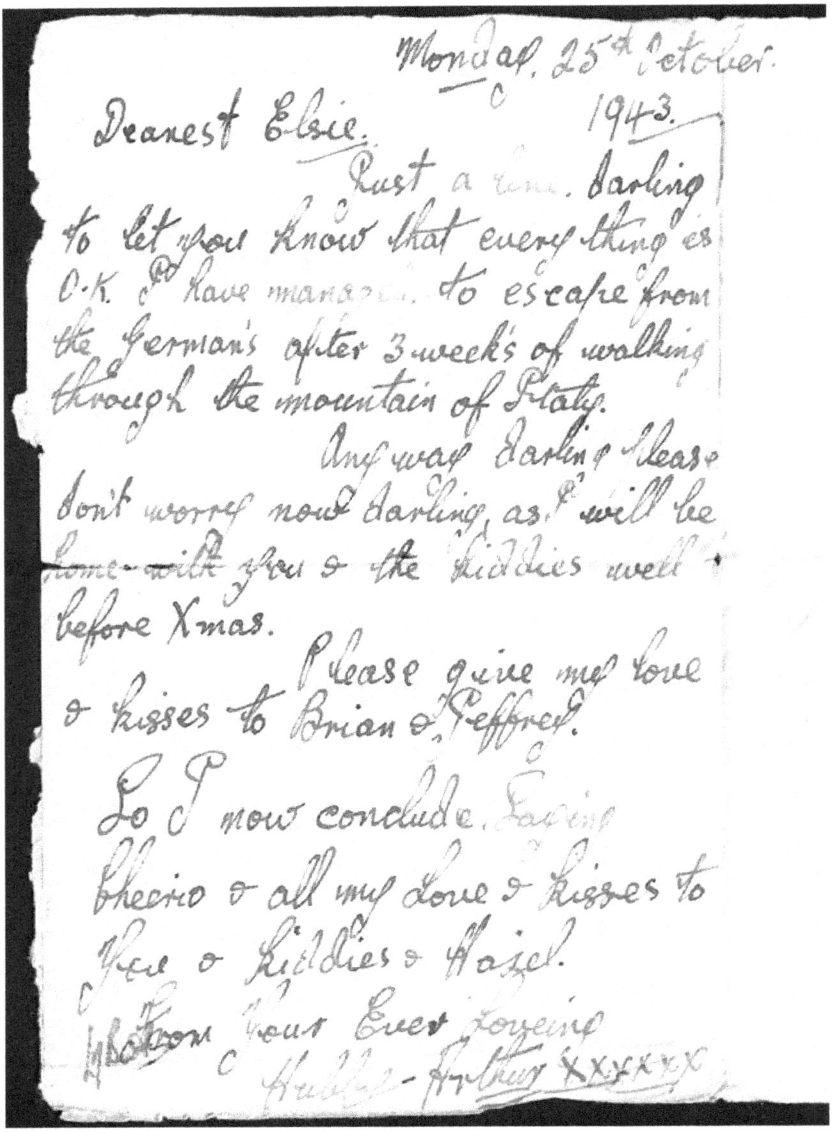

Monday. 25th October 1943.

Dearest Elsie.

Just a line. darling to let you know that every thing is O.K. I have managed to escape from the German's after 3 weeks of walking through the mountain of Italy.

Any way darling Please don't worry now darling, as I will be home with you & the kiddies well before Xmas.

Please give my love & kisses to Brian & Geffrey.

So I now conclude. Saying Cheerio & all my Love & kisses to You & kiddies & Hazel.

From Your Ever Loving Hubby - Arthur XXXXXX

Monday 25th October 1943

Dearest Elsie

Just a line darling to let you know that everything is OK. I have managed to escape from the Germans after 3 weeks of walking through the mountains of Italy.

Anyway, darling please don't worry now darling as I will be home with you and the kiddies well before Xmas.

Please give my love and kisses to Brian and Jeffrey.

So, I now conclude darling.

Cheerio and all my love and kisses to you and kiddies and Hazel

From your ever-loving Hubby

Arthur xxxxxx

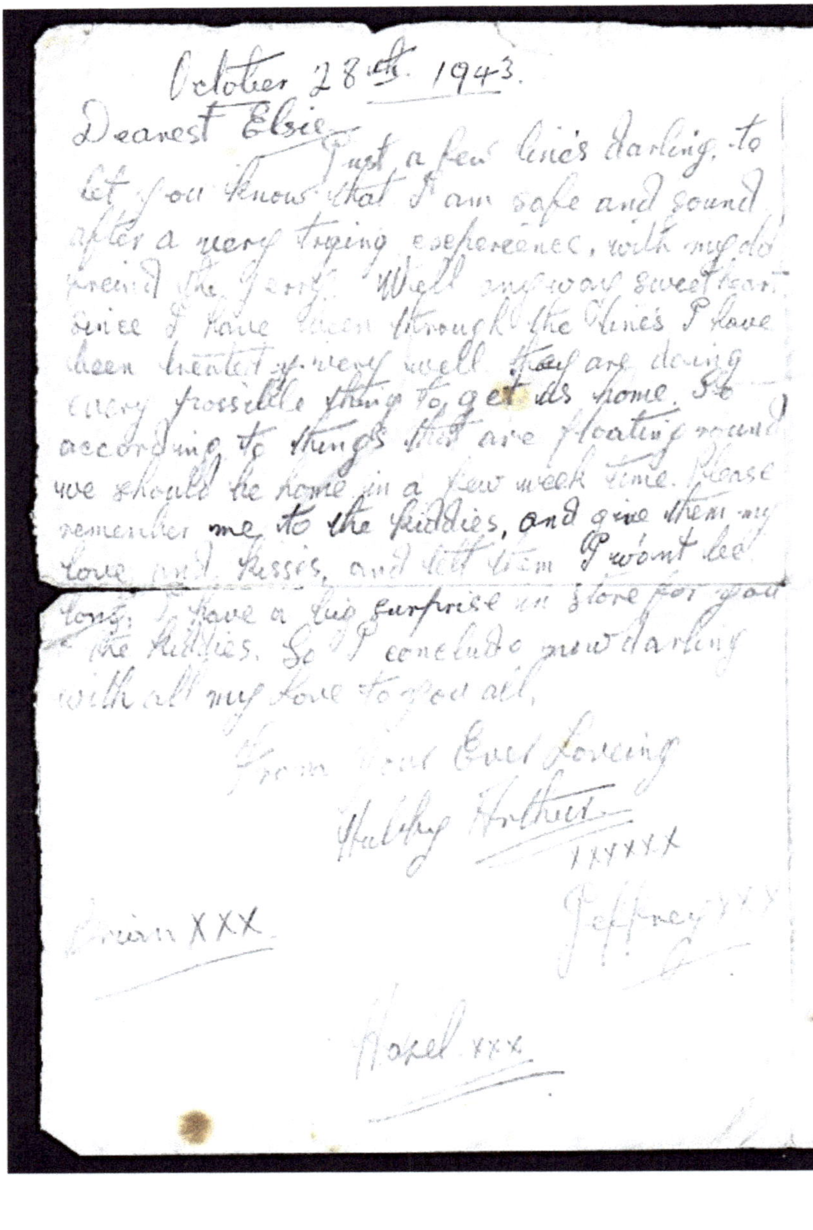

October 28th. 1943.

Dearest Elsie,
Just a few line's darling to let you know that I am safe and sound after a very trying expereence, with my old freind the Jerry. Well anyway sweetheart since I have been through the line's I have been treated very well they are doing every possible thing to get us home. So according to things that are floating round we should be home in a few week time. Please remember me to the kiddies, and give them my love and kisses, and tell them I won't be long. I have a big surprise in store for you & the kiddies. So I conclude now darling with all my love to you all,

From Your Ever Loving
Hubby Arthur
XXXXXX

Brian XXX

Jeffrey XXX

Hazel XXX

October 28th 1943

Dearest Elsie

Just a few lines darling to let you know that I am safe and sound after a very trying experience, with my old friend Jerry. Well anyway sweetheart, since I have been through the lines I have been treated very well. They are doing every possible thing to get us home. So according to things that are floating round we should be home in a few week time. Please remember me to the kiddies and give them my love and kisses and tell them I won't be long. I have a big surprise in store for you and the kiddies, so I conclude now darling with all my love to you all.

From your ever loving hubby Arthur xxxxxx

Brian xxx

Jeffrey xxx

My grandad carried his memories with him: the escape in Tobruk, Campo 65, transfer to Campo 145, and every road between captivity and freedom.

And now, so do I.

As I stood looking at the mountains he walked through, I suddenly thought he was a hero, something I had never thought before.

I used to remember him as my scruffy grandad with dirty hands… but now he is my hero

I hear his voice in my mind, quiet, steady, unshaken, just as it must have sounded to those who looked to him for strength:

"One step at a time. One day at a time. That's how you survive."

Approximate location of Campo 145

War monument at Montorio al Vomano

The mountains my Grandad walked through while escaping

CHAPTER EIGHT

LETTERS FROM CAMPO 145 AND ESCAPE

These are the letters from Campo 145, which, from what I have been told, are very rare, and there are not many that have survived.

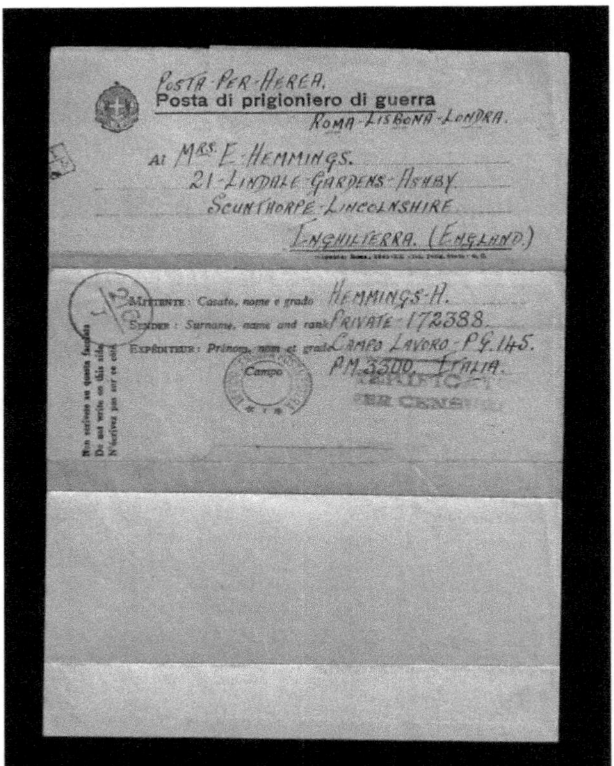

POSTA PER AEREA.
Posta di prigioniero di guerra
ROMA-LISBONA-LONDRA.

AI MRS. E. HEMMINGS.
21 LINDALE GARDENS ASHBY
SCUNTHORPE LINCOLNSHIRE
INGHILTERRA. (ENGLAND.)

MITTENTE: Casato, nome e grado HEMMINGS H.
SENDER: Surname, name and rank PRIVATE 172388.
EXPÉDITEUR: Prénom, nom et grade CAMPO LAVORO P.G. 145.
P.M. 3300. ITALIA.

Non scrivere su questa facciata
Do not write on this side
N'écrivez pas sur ce côté

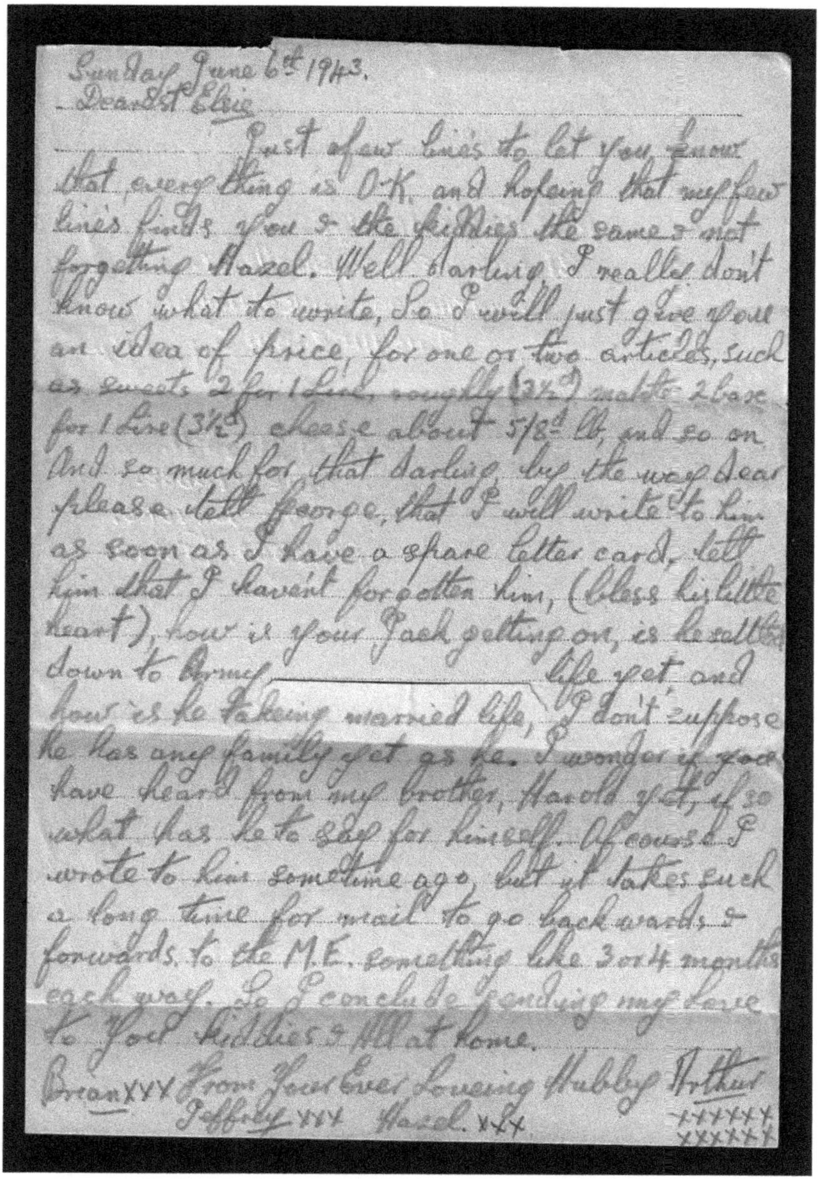

Sunday June 6th 1943.
Dearest Elsie
Just a few lines to let you know that every thing is O.K. and hopeing that my few lines finds you & the kiddies the same & not forgetting Hazel. Well darling I really don't know what to write, So I will just give you an idea of price, for one or two articles, such as sweets 2 for 1 lira, roughly (3½d) matches 2 box for 1 lira (3½d) cheese about 5/8d lb, and so on. And so much for that darling, by the way dear please tell George, that I will write to him as soon as I have a spare letter card, tell him that I haven't forgotten him, (bless his little heart), how is your Jack getting on, is he settled down to Army life yet, and how is he takeing married life, I don't suppose he has any family yet as he. I wonder if you have heard from my brother, Harold yet, if so what has he to say for himself. Of course I wrote to him sometime ago, but it takes such a long time for mail to go backwards & forwards to the M.E. something like 3 or 4 months each way. So I conclude sending my love to You Kiddies & All at home.
Brian XXX From Your Ever Loveing Hubby Arthur
Jeffrey XXX Hazel. XXX
XXXXX
XXXXX

Sunday June 6th 1943

Just a few lines to let you know that everything is OK and hoping that my few lines find you and the kiddies the same and not forgetting Hazel. Well darling I really don't know what to write, so I will just give you an idea of price, for one or two articles, such as sweets 2 for 1 roughly (3 ½ d), matches 2 base for 1 line (3 ½ d), cheese about 5/8 d lb and so on and so much for that darling. By the way dear please tell George that I will write to him as soon as I have a spare letter card. Tell him that I haven't forgotten him (bless his little heart). How is your Jack getting on, is he settled down into Army life, and how is h taking married life. I don't suppose he has any family yet has he. I wondered if you have heard from my brother Harold yet, is so what has he to say for himself. Of course I wrote to him some time ago but it takes such a long time for mail to go backwards and forwards to the M.E. something like 3 or 4 month each way. So, I conclude sending my love to you, kiddies and all at home

From your ever loving hubby Arthur xxxxxxxxxxxx

Brian xxx

Jeffrey xxx

Hazel xxx

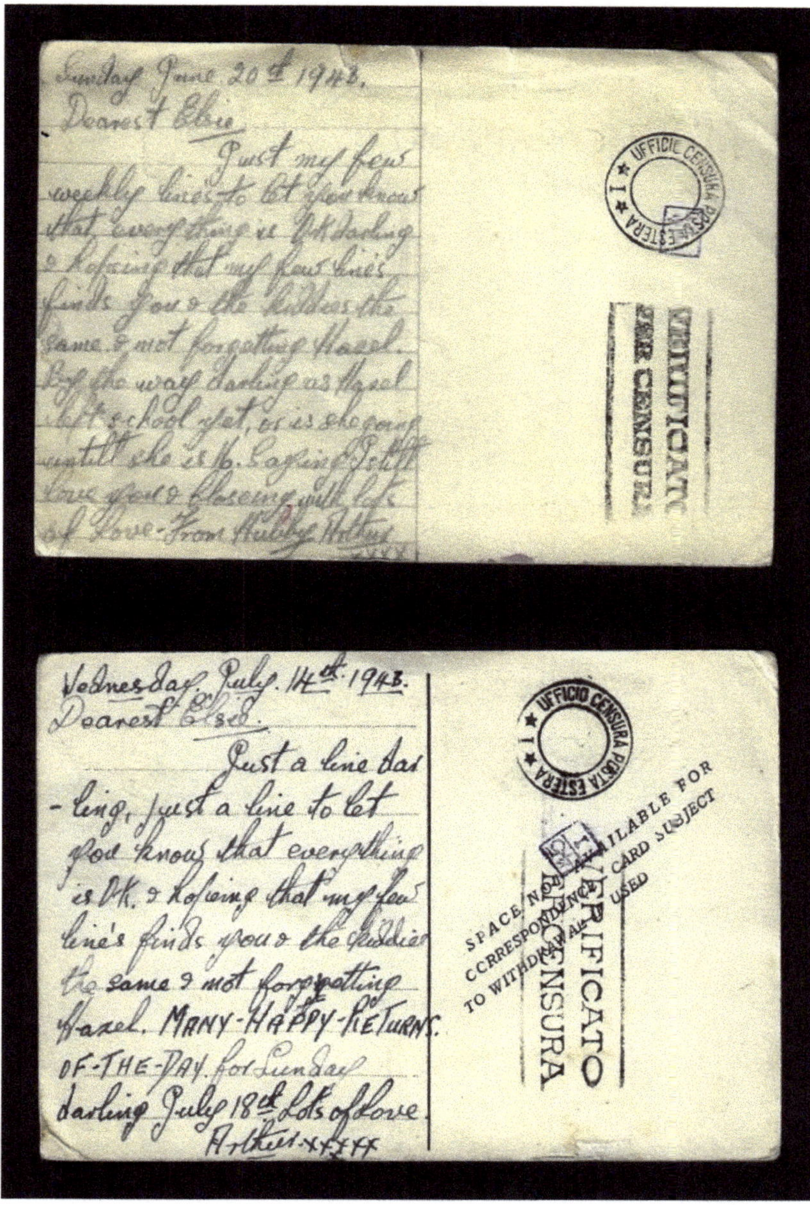

Sunday June 20th 1943.
Dearest Elsie
Just my few weekly lines to let you know that everything is OK darling & hoping that my few line's finds you & the kiddies the same & not forgetting Hazel. By the way darling as Hazel left school yet, or is she going until she is 16. Saying I still love you & closing with lots of love - From Hubby Arthur xxxx

UFFICIO CENSURA POSTA ESTERA I
VERIFICATO PER CENSURA

Wednesday July 14th 1943.
Dearest Elsie.
Just a line dar-ling, just a line to let you know that everything is OK. & hoping that my few line's finds you & the kiddies the same & not forgetting Hazel. MANY-HAPPY-RETURNS. OF-THE-DAY for Sunday darling July 18th Lots of love
Arthur xxxx

UFFICIO CENSURA POSTA ESTERA I
SPACE NOT AVAILABLE FOR CORRESPONDENCE CARD SUBJECT TO WITHDRAWAL IF USED
VERIFICATO PER CENSURA

Sunday June 20th 1943

Dearest Elsie

Just my few weekly lines to let you know that everything is ok darling, hoping that my few lines finds you are the kiddies the same and not forgetting Hazel. By the way darling has Hazel left school yet or is she going until she is 16. Saying I still love you and closing with lots of love from hubby Arthur xxxx

Wednesday July 4th 1943

Dearest Elsie

Just a line daring, just say line to let you know that everything is OK and hoping that my few lines finds you and the kiddies the same and not forgetting Hazel. MANY HAPPY RETURNS OF THE DAY for Sunday darling July 18th. Lots of love Arthur xxxxx

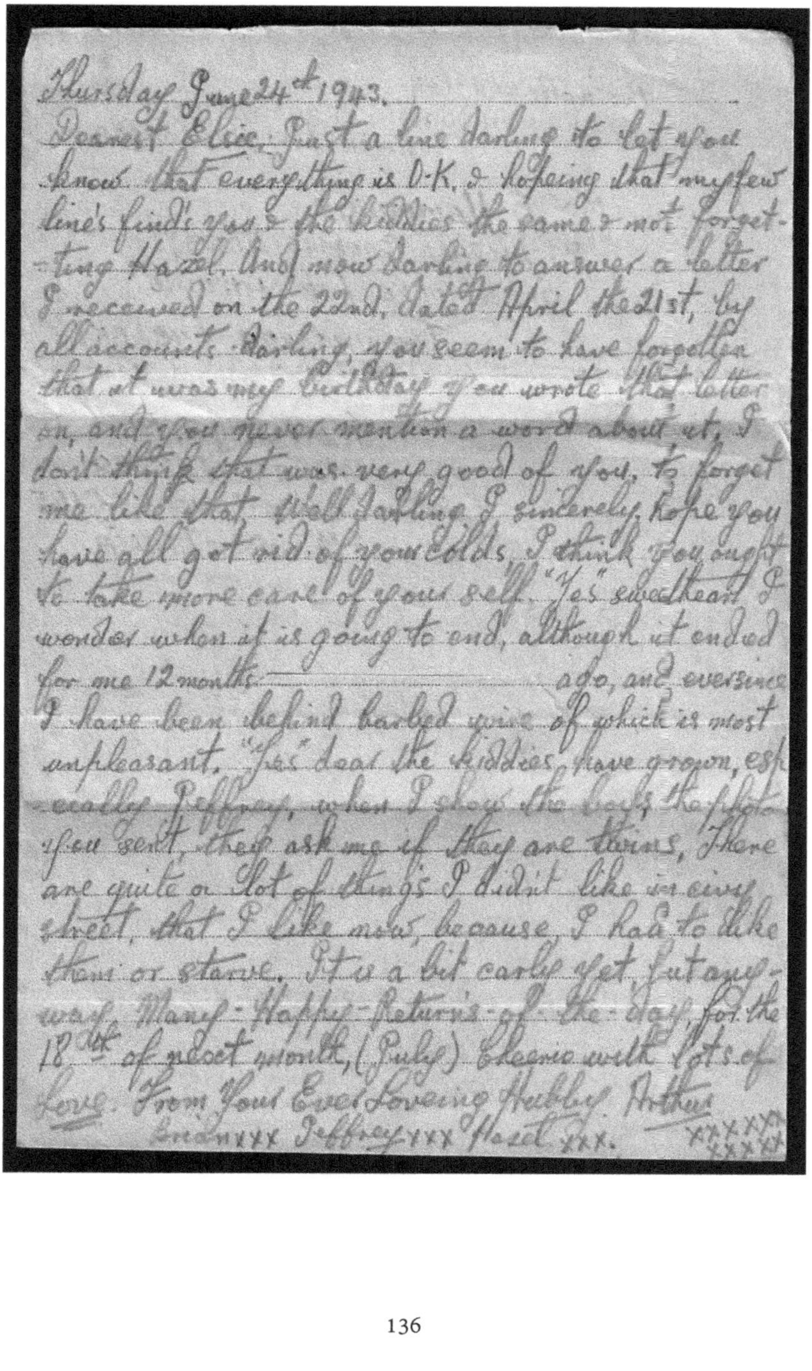

Thursday June 24th 1943.

Dearest Elsie, Just a line darling to let you know that everything is O.K. & hopeing that my few line's find's you & the kiddies the same & not forgetting Hazel. And now darling to answer a letter I received on the 22nd, dated April the 21st, by all accounts darling, you seem to have forgotten that it was my birthday you wrote that letter on, and you never mention a word about it, I don't think that was very good of you, to forget me like that, Well darling I sincerely hope you have all got rid of your colds, I think you ought to take more care of your self. "Yes" sweetheart I wonder when it is going to end, although it ended for me 12 months ago, and eversince I have been behind barbed wire of which is most unpleasant, "Yes" dear the kiddies have grown, especially Jeffrey, when I show the boys the photo you sent, they ask me if they are twins, There are quite a lot of things I didn't like in civy street, that I like now, because I had to like them or starve. It is a bit early yet, but anyway. Many - Happy - Return's - of - the - day, for the 18th of next month, (July) Cheerio with lots of love. From Your Ever Loving Hubby. Arthur

Brian xxx Jeffrey xxx Hazel xxx. xxxxx xxxxx

Thursday June 24th 1943

Dearest Elsie,

Just a line daring to let you know that everything is OK and hoping that my few lines find you and the kiddies the same and not forgetting Hazel. And now darling to answer a letter I received on the 22nd dated April the 21st, by all accounts darling you seem to have forgotten that it was my birthday you wrote that letter on and you never mentioned a word about it. I don't think that was very good of you to forget me like that. Well darling I sincerely hope you have all got rid of your colds. I think you ought to take more care of yourself. 'Yes' sweetheart I wonder when it is going to end, although it ended for me 1 months ag and ever since I have been behind barbed wire of which is most unpleasant. "Yes" dear the kiddies have grown especially Jeffrey when I show the photo you sent they ask me if they are twins. There are quite a lot of things I didn't like in civvies street that I like now because I have to like them or starve. It is a bit early yet but anyway Many Happy Returns of the day for the 18th of next month (July). Cheerio with lots of love from your ever loving Hubby Arthur xxxxxxxxxxx

Brian xxx

Jeffrey xxx

Hazel xxx

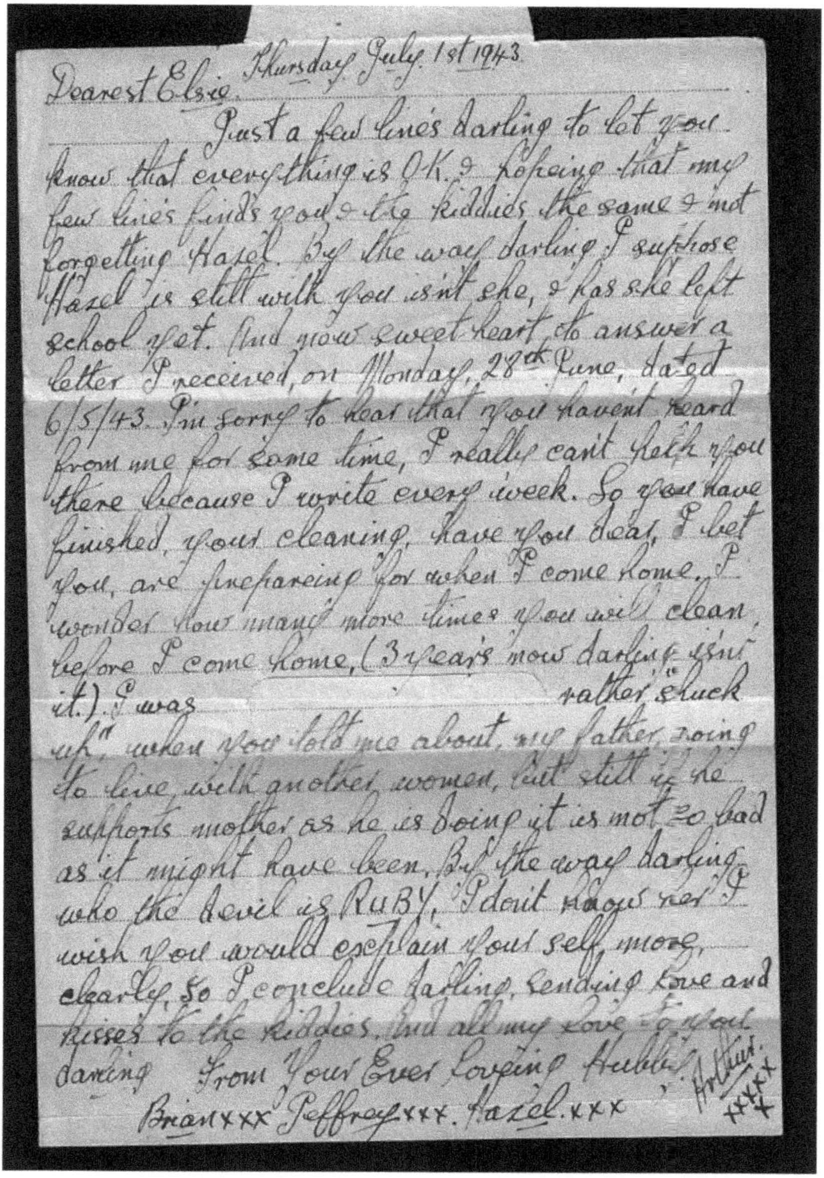

Dearest Elsie Thursday July 1st 1943.

Just a few line's darling to let you know that everything is O.K. & hopeing that my few line's finds you & the kiddies the same & not forgetting Hazel. By the way darling I suppose Hazel is still with you isn't she, & has she left school yet. And now sweet heart to answer a letter I received, on Monday, 28th June, dated 6/5/43. I'm sorry to hear that you haven't heard from me for some time, I really can't help you there because I write every week. So you have finished your cleaning, have you dear, I bet you are prepareing for when I come home. I wonder how many more times you will clean, before I come home, (3 year's now darling isn't it.) I was rather "shuck up" when you told me about my father going to live with another women, but still if he supports mother as he is doing it is not so bad as it might have been. By the way darling who the devil is RUBY, I don't know her I wish you would explain your self more clearly. So I conclude darling, sending love and kisses to the kiddies. And all my love to you darling from Your Ever Loving Hubby. Arthur. xxxxx

Brian xxx Jeffrey xxx. Hazel. xxx

Thursday July 1st 1943

Dearest Elsie,

Just a few lines daring to let you know that everything is OK. Hoping my few lines finds you and the kiddies the same ad not forgetting Hazel. By the way darling, I suppose Hazel is still with you isn't she and has she left school yet. And now sweetheart to answer a letter I received on Monday 28th June dated 06/05/43, I'm sorry to hear that you haven't heard from me for some time, I really can't help you there because I write every week. So you have finished cleaning have you dear, I bet you are preparing for when I come home. I wonder how many times you will clean before I come home (3 years now darling isn't it). I was rather shuck up when you told me about my father going to live with another woman but still if he supports mother as he is doing it is not so bad as it might have been. By the way darling, who the devil is RUBY, I don't know her wish you would explain yourself more clearly. So, I conclude darling sending love and kisses to the kiddies and all my love to you darling from your ever loving Hubby Arthur xxxxxx

Brian xxx Jeffrey xxx Hazel xxx

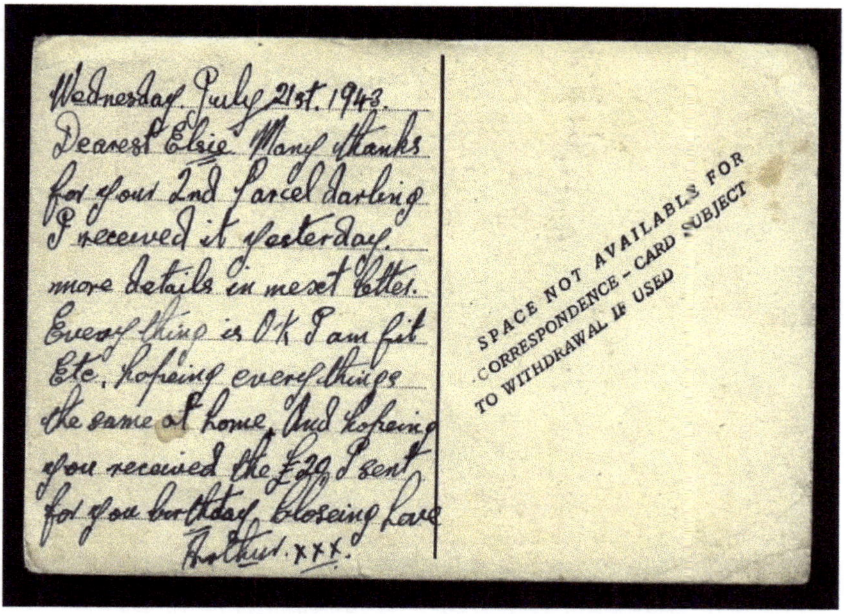
Wednesday. July 21st. 1943.
Dearest Elsie Many thanks
for your 2nd parcel darling
I received it yesterday.
more details in next letter.
Every thing is Ok I am fit
Etc. hopeing every things
the same at home. And hopeing
you received the £20 I sent
for you birthday. closeing love
Arthur. xxx.

SPACE NOT AVAILABLE FOR
CORRESPONDENCE - CARD SUBJECT
TO WITHDRAWAL IF USED

Wednesday July 12st 1943

Dearest Elsie,

Many thank for your 2nd parcel darling I received it yesterday. More details in next letter. Everything is Ok I am fit etc, hoping everything is the same at home and hoping you received the £20 I sent for your birthday. Closing love Arthur xxx

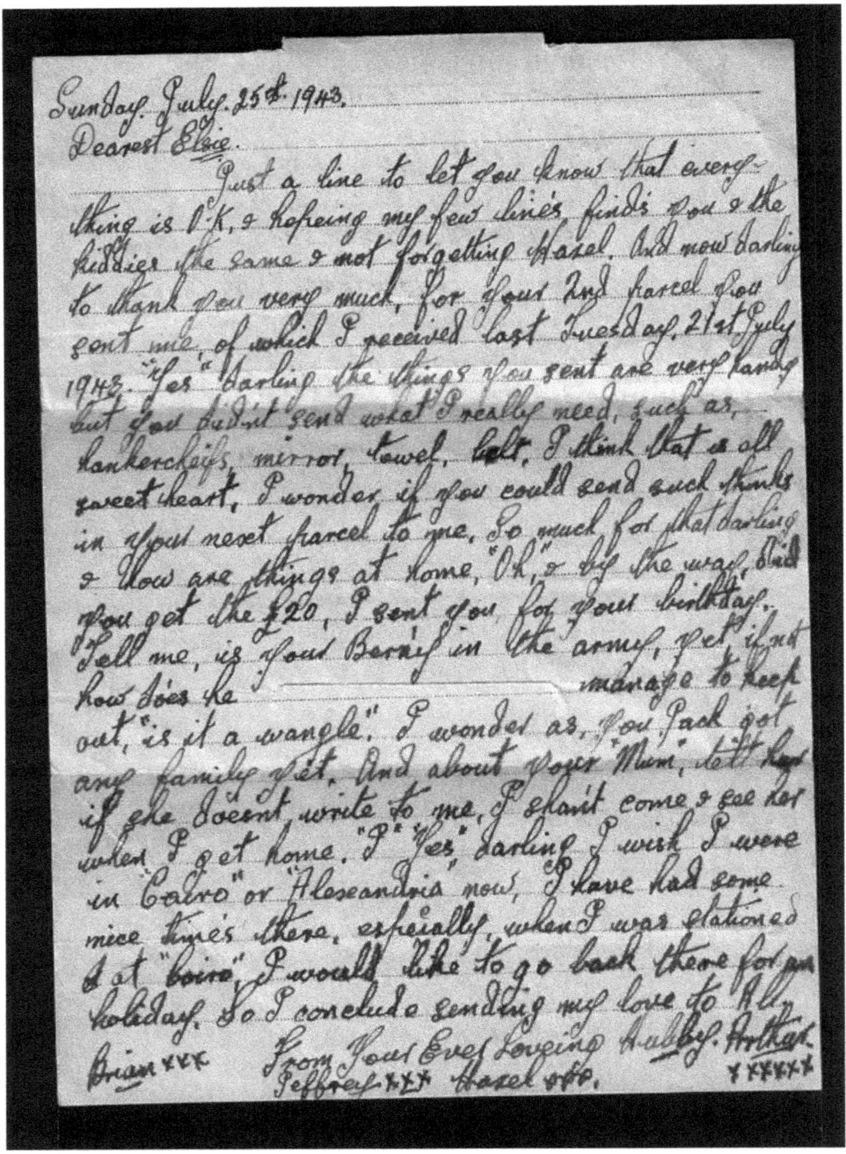

Sunday. July. 25th. 1943.
Dearest Elsie.
Just a line to let you know that every-
thing is O.K, & hopeing my few line's finds you & the
kiddies the same & not forgetting Hazel. And now darling
to thank you very much, for your 2nd parcel you
sent me, of which I received last Tuesday, 21st July
1943. "Yes" darling the things you sent are very handy
but you didn't send what I really need, such as,
hankercheifs, mirror, towel, belt, I think that is all
sweetheart, I wonder if you could send such thinks
in your next parcel to me. So much for that darling
& how are things at home, "Ok", & by the way, did
you get the £20, I sent you, for your birthday.
Tell me, is your Berni in the army, yet if not
how does he manage to keep
out, "is it a wangle". I wonder as, you Jack got
any family yet. And about your Mum, tell her
if she doesent write to me, I shaint come & see her
when I get home. "I" "Yes" darling I wish I were
in "Cairo" or "Alexandria" now, I have had some
nice time's there, especially when I was stationed
at "Cairo" I would like to go back there for a
holiday. So I conclude sending my love to All.
Brian xxx From Your Ever Loving Hubby. Arthur.
Jeffrey xxx Hazel xxx. xxxxxx

Sunday July 25th 1943

Dearest Elsie,

Just a line to let you know that everything is OK and hoping my few lines finds you and the kiddies the same and not forgetting Hazel. And now darling to thank you very much for your 2nd parcel you sent me of which I received last Tuesday 21st July 1943. 'Yes' darling the things you send are very handy, but you didn't send what I really needed such as handkerchiefs, mirror, towel, soap. I think that is all sweetheart, I wonder if you could send such thinks in your next parcel to me. So much for that darling and how are things at home, 'OK' and by the way did you get the £20 I sent you for your birthday. Tell me is your Barry in the army yet if not, how does he manage to keep out is it a wangle. I wonder has your Jack got any family yet. And about your Mum tell her if she doesn't write to me I shan't come and see her when I get home. Yes darling I wish I were in Cairo or Alexandria now. I have some nice times there especially when I was stationed at Cairo. I would like to go back there for a holiday. So I conclude and sending my love to you all.

From your ever-loving Hubby Arthur xxxxxx

Brian xxx

Jeffrey xxx

Hazel xxx

CHAPTER NINE

HIS MEDALS

On 25 April 1978, my grandad was finally claimed his Second World War medals: the 1939–1945 Star, the Africa Star, the War Medal 1939–45, and the Army Medal.

The date might seem surprising, coming more than thirty years after the war had ended, but this delay was far from unusual.

After the war, Britain did not automatically issue medals to every serviceman. Instead, soldiers were expected to apply for them.

Many never did.

Men returned home, found work, raised families, and tried to put the war behind them.

For former prisoners of war, especially, there was often little appetite for paperwork or revisiting painful memories.

Some didn't know they had to apply at all. Others simply felt that survival was enough and medals did not matter.

In many cases, medals were only claimed decades later, often prompted by family members, anniversaries, or changes in how the war was remembered.

By the 1970s, there was renewed public interest in recognising veterans properly, and many men, or their families, finally applied for the medals they were entitled to.

That is why my grandad's medals were issued in 1973, rather than immediately after the war.

Each medal tells part of his story.

The **1939–1945 Star** was awarded to members of the armed forces who served during the Second World War. It recognised operational service anywhere between September 1939 and September 1945. This medal marked his participation in the war itself, regardless of theatre, and acknowledged that he had served during a global conflict that defined a generation.

The **Africa Star** was awarded for service in North Africa between 1940 and 1943. This medal reflects my grandad's time in the desert war in Egypt, including the campaigns that led to the fall of Tobruk and his capture. It represents service under extreme conditions, heat, thirst, constant movement, and relentless fighting, and ties directly to the stories he later told about escaping into the desert and struggling to survive.

The **War Medal 1939–45** was awarded to all full-time personnel who completed at least twenty-eight days of service during the war. It was one of the most widely issued medals, recognising the collective effort of those who served, whether in combat, captivity, or support roles. For former prisoners of war, it acknowledged endurance as much as action.

The **Italy Star Medal** was awarded to soldiers of the British and Commonwealth forces who served in the Italian campaign during the

Second World War. To receive it, personnel had to complete at least one day of operational service in Italy or surrounding areas between 1943 and 1945, often under challenging and dangerous conditions. The medal recognised the terrain, harsh weather, and determined resistance faced during campaigns such as the advance through Sicily and mainland Italy. It served as a mark of honour for those who contributed to the Allied effort in liberating Italy from Axis control.

Taken together, these medals tell a fuller story than any single one could.

They speak of enlistment and service, of desert warfare, of capture and captivity, and of endurance.

Even though they were awarded so many years later does not diminish their meaning.

If anything, it highlights how men like my grandad carried their war quietly, without expectation of recognition, until someone finally took the time to ask the questions and fill in the forms.

I have all of my grandad's medals.

Holding them now, knowing what he endured to earn each one, feels very different from simply seeing them mounted in a picture frame.

They are small, solid things, worn smooth with time, yet they carry the weight of deserts, camps, hunger, fear, and survival.

My dad inherited his medals when he died, and I inherited them from him.

I will cherish these medals and will ensure they are passed down the family tree so no one will ever forget him.

Photo Of Medals

CHAPTER TEN

A LIFE LESS ORDINARY

After the war, my grandad did what many men did. He came home and tried to build something steady.

Arthur and Elsie's family grew. Allan, my dad, was born after his return at the end of December 1944.

Then Graham, who sadly died young, less than a year old.

Two daughters followed, Eileen in 1946 and Janet in 1950.

There is also mention of another child, Sheila, born outside of marriage, in July 1946. The details are uncertain, spoken quietly rather than openly, but she too is part of this story.

For a time, he did well.

He had a garage and a lovely bungalow.

On the surface, it must have looked as though life had settled into place.

A business of his own. A roof over his head. A family growing around him.

After everything he had been through, the desert, captivity, escape, perhaps he wanted something solid. Something that belonged to him.

But my grandad was never a man who liked rules, especially not official ones.

At some point, he stopped paying taxes.

Whether it was oversight, defiance, or simply his way of doing things, I don't know.

What I do know is that HMRC eventually caught up with him.

His business was taken. The bungalow was taken.

For some men, that might have broken them.

For my grandad, it became the beginning of something else.

He bought a double-decker bus.

Not to run as transport. Not to scrap. But to convert into a family home.

And that is exactly what he did. He transformed it into somewhere to live and parked it in a field in Brigg. There it stayed for two years.

To him, I imagine it felt practical. Resourceful. A solution to a problem.

He had survived a war, after all. Living in a converted bus probably didn't seem so extreme.

But when I recently spoke to my Auntie Eileen, Authur's daughter, she told me it was the worst two years of her life.

I can picture it now.

A red bus standing alone in a field.

Wind across open land. Cold mornings. Limited space. No privacy.

It may have been an adventure in theory, but in reality, it was simply hard. Especially for children who wanted normality, not novelty.

That was my grandad. Independent. Stubborn. Creative.

Unconcerned with what other people thought. He would rather build his own solution than follow someone else's path.

Things changed, and life got better.

When I was born, in January 1969, my nana and grandad owned a corner shop on Diana Street in Scunthorpe.

I remember staying from time to time, and my grandad used to let us have one toy from the shop, now and again.

I also have a memory of him taking my sister and me to the cash and carry. I stood on the trolly which was long and only had one wheel in the middle. I used to love it.

Then, at some point that was gone too, I think this was when my nana and grandad split up and went their separate ways.

By the early 1970s, the strain showed. My nana and grandad divorced.

They had weathered war, separation, and rebuilding, but peace brought its own challenges. Both of them remarried and went on to build different lives.

Years later, my nana would sometimes say, with a certain sharpness, that my grandad had left her for four years with two children.

She used to say it as if it were a recent occurrence.

I would gently remind her that he had been a prisoner of war, that he hadn't exactly chosen to stay away.

Not because it was funny, really, but because it showed how time softens and reshapes memory.

To her, those years must have felt like abandonment. Even though he had been a POW and couldn't come home, the hurt remained.

The loneliness. The uncertainty.

Raising two boys, not knowing whether her husband was alive or dead.

It reminded me that war does not end when the fighting stops. It carries on inside marriages, inside families, inside quiet resentments and unspoken pain.

My own memories of my grandad are ordinary ones, and because of that, precious.

I used to cut his hair every four weeks.

A chair in the kitchen or living room, scissors and clippers in my hands. It was our quiet routine.

He trusted me, and that trust mattered. He used to sit with a bottle of alcoholic free beer, due to having diabetes, while I cut his hair.

He was not polished or showy.

My dad, Allan, was sometimes embarrassed by him.

Grandad never had much money and never cared about appearances.

I remember him turning up once in an old ambulance he had converted into a campervan.

To him, it was practical.

To my dad, mortifying.

Looking back now, I admire him.

He did not worry about what people thought. He made things work.

He spoke little about the war.

What he had been through stayed mostly inside him.

My grandad was many things. A mechanic. A soldier. A prisoner. An escapee. A businessman. A dreamer who turned buses into homes. A man who sometimes got things wrong.

But he was never ordinary.

And perhaps that is the truest way to remember him.

By the late 1980's, my granddad's health deteriorated.

He was taken to the hospital in 1985 with heart, liver, and lung failure, and again in December 1988. This time, he did not survive.

I visited him on the day he died.

I walked into his room, and he was sitting in a hospital chair at the end of his bed.

He had a glow about him that I hadn't seen before. I had never seen him this clean.

As I have mentioned before, his hands were always dirty.

As he was sitting in the chair, he had on a new pair of pyjamas that were pale blue and white striped. He had his thumb in the elastic.

'What is wrong grandad?" I asked.

He just made a cutting motion.

'Is the elastic too tight?" thumbs up.

'Do you want me to get some scissors to cut it?' thumbs up again.

It was like he had lost his voice.

After the elastic was cut, he relaxed, and I just chatted away at him, and he listened.

I left about an hour later, and that was the last time I saw him.

It was a sad day.

But I can now truly say he died not only as my grandad, but also as a true war hero.

"One day at a time".

CHAPTER ELEVEN

WE FOLLOWED IN HIS FOOTSTEPS.

War did not end with my grandad.

Four of the Hemmings clan followed in my grandad's footsteps.

First, his two sons: Brian, who served for nine years in the Royal Ordnance Corps, and Allan, my dad, who served in the Royal Anglian Regiment and the Royal Army Medical Corps (RAMC).

Secondly, his eldest grandson, Pete, Brian's son, served in the RAMC, followed by me, his granddaughter, Allan's daughter, who served in the Signals before moving to the Royal Logistic Corps (RLC).

But I want to tell you a bit about my dad.

My dad was Allan Michael Hemmings, born on 14th December 1944, 11 months after my grandad returned home from the war.

To most people, he was just Allan.

But in the Army, he was known as Major H, which stood for Major Hairspray, due to his immaculate hair, which never seemed to move, no matter the weather, the helmet, or the situation.

It was a nickname given with affection, and it stuck.

He joined the Army as a young man, following the same path his father had taken years before.

He enlisted as a private with the 1st Royal Anglian Regiment, better known as the Pompadours.

Like many before him, he started at the bottom of the ranks as a Private. There were no shortcuts. Rank was earned, not assumed.

His service took him abroad.

He was posted to Malta, a place steeped in its own wartime history, still carrying the scars of siege and bombardment.

From there, he served in Germany, a country still living in the long shadow of the war my grandad had fought.

One of the most extraordinary chapters of my dad's service came during his time in Berlin.

He was stationed at Spandau Prison.

This was where he guarded some of the most infamous men of the Second World War.

Rudolf Hess, Albert Speer and Baldur von Schirach.

These were names tied to unimaginable destruction, now reduced to prisoners behind walls, guarded by soldiers like my dad.

I heard a story recently that these men tried to bride my dad due to his blonde hair and blue eyes, the perfect German combination.

It is strange to think of the connection.

My grandad had been captured by Italian forces, imprisoned, starved, and forced to escape to survive.

Years later, his son stood guard over the defeated architects of that same war.

History has a way of folding back on itself like that.

My dad never spoke much about Spandau.

Like his father, he didn't dwell on the weight of it. It was a job to be done, and he did it.

He left the regular Army when he met his wife, my mum, Ann Bird.

Life pulled him in a different direction.

He paid to leave, choosing love and family over a full-time military career.

But the Army never fully let him go, and perhaps he never wanted it to.

He joined the Territorial Army in Scunthorpe, based at Cottage Beck Road, where the Royal Anglian Regiment was also stationed.

He had only spent four years in the regular Army, joining the TA as a Private soldier. Over time, through commitment and sheer persistence, he worked his way up the ranks.

Private became corporal. Corporal became an officer. The officer became Major.

After many years in the Royal Anglian Regiment, he moved to the Royal Medical Corps at Wenlock Barracks in Hull, where, for a time, he even served as Acting Colonel.

In total, he gave twenty-nine years of service. Nearly three decades of weekends, evenings, exercises, camps, responsibility, and leadership, all balanced alongside civilian life.

During those years, he was recognised for his service.

He was invited to the Queen's Garden Party, a moment of quiet pride.

He was also awarded the Lord Lieutenant's Certificate, another mark of respect for long and dedicated service. These weren't things he boasted about. They were accepted with the same calm professionalism that defined him.

Growing up, the Army was always there in the background. Stories overheard rather than told. Names and ranks that became familiar without explanation. It felt expected, looking back, that I would follow too.

I was nineteen when I joined the Territorial Army.

I started in the Signals at Hull, learning radio communication, structure, and discipline.

Later, I transferred to the Royal Logistic Corps, attached to the Royal Electrical and Mechanical Engineers (REME) in Scunthorpe. By then, I had been promoted to Lance Corporal and found myself stationed at the very same Cottage Beck Road base, where my dad had once served and my grandad had enlisted.

Some of the commissioned officers and NCOs I then worked with had served with my dad in the Royal Anglians.

In that place, I wasn't just myself. I was "H's daughter." I grew up in these barracks and knew every room, every nook and cranny. I loved going there with my dad.

Standing there in uniform, I felt the weight of something bigger than me.

Three generations connected by service.

My grandad, a prisoner of war who escaped camps and crossed mountains.

My dad was the soldier who guarded Germany's war criminals and rose through the ranks.

My Uncle Brian, cousin Pete, and I, stepping into that long shadow of my Grandad.

We didn't all serve in the same way.

Different wars. Different roles. Different times.

But the thread is there, unbroken.

When I look back now, I see that my grandad's silence did not end his story. It carried forward. It shaped my dad. And through him, it shaped me.

We followed in his footsteps, not because we were told to, but because that was simply the path laid before us.

And now, by writing this, I take one final step along it.

EPILOGUE

It wasn't until I was flying back home, after visiting the camps, that it truly settled inside me.

I had carried the letters back with me, tucked carefully into my bag, the same letters my grandad had written all those years ago.

I had brought them to Italy as a reference, as evidence, as something solid to hold onto while I walked the places where he had once been held.

But it was only on the journey home that I really understood what they meant.

The plane lifted into the air, Italy shrinking beneath the clouds, the mountains fading into the distance.

Somewhere below me were Campo 65 and Campo 145.

The deserts. The huts.

The hills my grandad had crossed on foot.

Places that had once defined his world and now sat quietly in mine.

I stared out of the window, thinking about how easily the land hides what happened on it.

Fields return to pasture. Buildings fall. Camps disappear. Only stories remain.

I took the letters out one by one.

His handwriting was familiar, steady, and careful.

There was no drama in his words. No anger. No self-pity.

He asked after my nana. He spoke of the weather. He reassured her that he was fine, that she shouldn't worry, that he would be home.

Reading them now, with the knowledge of where he was when he wrote them, was almost unbearable.

I now know what he was not saying. I now know what those calm sentences were not saying.

He never described the hunger in detail. He never spoke of the fear. He never wrote about the swamps, the sickness, the punishment cells, or the escape through the mountains.

He protected the people he loved by staying quiet. Silence, I realised, was one of the ways he survived.

As the plane hummed steadily around me, I thought about how long this story had waited to be told.

Decades, really.

Passed down in fragments.

Half-remembered conversations.

A child listening wide-eyed to a tale about escaping into the desert and drinking his urine to survive, without understanding what that truly meant.

A family that knew him as a husband, a father, a grandfather, but not fully as the man shaped by war.

I thought of my dad, following him into service. Of myself doing the same.

Of how history moves through families quietly, settling into habits, choices, silences.

I thought about how close this story had come to being lost altogether. If the letters had been thrown away.

If no one had asked the questions. If I had never gone to Italy. If, If, If.....

Holding those letters, I felt a sense of responsibility I hadn't expected.

Not just to remember him, but to speak for him. To say the things, he never did. To place his life back into the history that shaped it.

This book is not about glory.

It is not about medals or heroics in the way films portray them.

It is about endurance.

About ordinary men placed in extraordinary circumstances and finding a way through.

It is about my grandad, Arthur Hemmings, a mechanic, a soldier, a prisoner of war, an escapee, a survivor.

When the plane began its descent, I put the letters back into my bag. I knew then that I wasn't finished. That the journey hadn't ended in Italy. It was only just beginning.

This book is my way of bringing him home.

You will never be forgotten.

'One day at a time'

APPENDICES

Military armistice signed at Fairfield Camp, Sicily, September 3, 1943

Entered into force September 3, 1943

Supplemented by memorandum of agreement of September 23, 1943, as amended, and by instrument of surrender of September 29, 1943, as amended

Terminated September 15, 1947, upon entry into force of the Treaty of Peace of February 10, 1947

61 Stat. 2740

Treaties and Other International Acts Series 1604

FAIRFIELD CAMP
SICILY
September 3, 1943

The following conditions of an Armistice are presented by

General Dwight D. Eisenhower,

Commander-in-Chief of the Allied Forces, acting by authority of the Governments of the United States and Great Britain and in the interest of the United Nations, and are accepted by

Marshal Pietro Badoglio, Head of the Italian Government

1. Immediate cessation of all hostile activity by the Italian armed forces.

2. Italy will use its best endeavours to deny, to the Germans, facilities that might be used against the United Nations.

3. All prisoners or internees of the United Nations are to be immediately turned over to the Allied Commander in Chief, and none of these may now or at any time be evacuated to Germany.

4. Immediate transfer of the Italian Fleet and Italian aircraft to such points as may be designated by the Allied Commander in Chief, with details of disarmament to be prescribed by him.

5. Italian merchant shipping may be requisitioned by the Allied Commander in Chief to meet the needs of his military-naval program.

6. Immediate surrender of Corsica and of all Italian territory, both islands and mainland, to the Allies, for such use as operational bases and other purposes as the Allies may see fit.

7. Immediate guarantee of the free use by the Allies of all airfields and naval ports in Italian territory, regardless of the rate of evacuation of the Italian territory by the German forces. These ports and fields are to be protected by Italian armed forces until this function is taken over by the Allies.

8. Immediate withdrawal to Italy of Italian armed forces from all participation in the current war from whatever areas in which they may be now engaged.

9. Guarantee by the Italian Government that if necessary, it will employ all its available armed forces to ensure prompt and exact compliance with all the provisions of this armistice.

10. The Commander in Chief of the Allied Forces reserves to himself the right to take any measure which in his opinion may be necessary for the protection of the interests of the Allied Forces for the prosecution of the war, and the Italian Government binds itself to take such administrative or other action as the Commander in Chief may require, and in particular the Commander in Chief will establish Allied Military Government over such parts of Italian territory as he may deem necessary in the military interests of the Allied Nations.

11. The Commander in Chief of the Allied Forces will have the full right to impose measures of disarmament, demobilization, and demilitarization.

12. Other conditions of a political, economic, and financial nature with which Italy will be bound to comply will be transmitted at a later date.

The conditions of the present Armistice will not be made public without prior approval of the Allied Commander in Chief. The English will be considered the official text.

MARSHAL PIETRO BADOGLIO
Head of the Italian Government

By:
GUISEPPE CASTEI.LANO
Brigadier General, attached to the Italian High Command

Present:
Rt. Hon. Harold Macmillan
British Resident Minister, A.F.H.Q.

Robert Murphy
Personal Representative of the
President of the United States

Royer Dick
Commodore, R.N.
Chief of Staff to the C. in C. Med.

DWIGHT D. EISENHOWER
General, U.S. Army,
Commander in Chief, Allied Forces

By:

WALTER B. SMITH
Major General, U.S. Army,
Chief of Staff

Lowell W. Rooks
Major General, U.S. Army
Assistant Chief of Staff, G-3
A.F.H.Q.

Franco Montanari
Official Italian Interpreter

Brigadier Kenneth Strong
Assistant Chief of Staff, G-3
A.F.H.Q.

Source:
Treaties and Other International Agreements of the United States of America 1776-1949
Compiled under the direction of Charles I. Bevans, LL.B.
Assistant Legal Advisor Department of State
Volume 3 Multilateral 1931-1945
Department of State Publication 8484
Washington, DC: Government Printing Office, 1969

www.ingramcontent.com/pod-product-compliance
Lightning Source LLC
LaVergne TN
LVHW050914210926
848482LV00012B/428
* 9 7 9 8 9 0 5 1 4 3 0 1 4 *